SAYDULLAYEV TOIRJON SAYDULLAYEVICH
KURBANOVA NILUFAR KUBIJANOVNA
DEHQONOVA DILOVAR KOMILJANOVNA

Modern methods of studying biology for medical students

SAYDULLAYEV TOIRJON SAYDULLAYEVICH
KURBANOVA NILUFAR KUBIJANOVNA
DEHQONOVA DILOVAR KOMILJANOVNA

Modern methods of studying biology for medical students

MONOGRAPH

LAP LAMBERT Academic Publishing

Imprint

Cover image: www.ingimage.com

Publisher:
LAP LAMBERT Academic Publishing
is a trademark of
Dodo Books Indian Ocean Ltd. and OmniScriptum S.R.L publishing group

120 High Road, East Finchley, London, N2 9ED, United Kingdom
Str. Armeneasca 28/1, office 1, Chisinau MD-2012, Republic of Moldova, Europe
Managing Directors: Ieva Konstantinova, Victoria Ursu
info@omniscriptum.com

Printed at: see last page
ISBN: 978-620-9-10577-7

MINISTRY OF HIGH EDUCATION, SCIENCE AND INNOVATIONS OF THE REPUBLIC OF UZBEKISTAN

ANDIJAN STATE MEDICAL INSTITUTE

SAYDULLAYEV TOIRJON SAYDULLAYEVICH, KURBANOVA NILUFAR KUBIJANOVNA, DEHQONOVA DILOVAR KOMILJANOVNA

MODERN METHODS OF STUDYING BIOLOGY FOR MEDICAL STUDENTS

MONOGRAPH

ANDIJAN 2025

SAYDULLAYEV TOIRJON SAYDULLAYEVICH, - *associated professor of the department of Histology and Medical Biologist*

NILUFAR KURBANOVA KUBIJANOVNA - Department of Medical Biology and Histology, Senior Lecturer

DEHQONOVA DILOVAR KOMILJANOVNA - Assistant Professor of the Department of Medical Biology and Histology

CONTENTS

ACCESS

To improve the quality of teaching of chemistry in the world, the introduction of innovation and information technology in the education process, integrasiyalash, mixed technology education opportunities for the use of intellectual potential of students, aimed at the development of creative skills and modern methods of cognition aimed at creating effective supply level of the studies being conducted. The world's leading research centers and higher education institutions of the science of chemistry by specialists in innovation and information technologies for the improvement of the training process on the basis of theoretical-methodological and methodical recommendations developed on the principles of scientific developments in the process of education to apply.

At the international level to increase the effectiveness of the teaching of natural sciences, chemistry, information technologies and innovative pedagogical technologies of formation of the theoretical basis of using paid particular attention to. At present, foreign countries, educational and scientific laboratories completed the practical work on the creation of a number. The experiences from the hard drive or the internet in this research work using two-dimensional graphic models is carried out using. Three-dimensional graphics in the virtual lab chemical chemical jamlanma developed for the experiment. Virtual chemistry at the university oksford videolavha designed to fulfill the experience created.

In the republic on the issues of improving the education system and harmoniously developed generation rose to the priority level of state policy. Education system of our country and the introduction of innovative technologies in the educational process in the information and communication tools, training intensity and to increase the effectiveness of them is being carried out extensive reforms to adapt to the world standards. The development of national education system of the republic of uzbekistan until 2030 kontsepsiyasida "the computer for the improvement of teaching, the principle of the gradual introduction of individuallashtirish to the process of education, training and innovation projects in the sphere of introduction of modern information and communication technologies" of the task set. In this regard, Iqtisodiy the conditions of synthesis of new chemical compounds and materials which would serve to increase the capacity to develop their composition, structure and texture, the study of actual problems of modern chemistry as applying in certain areas plays an important role.

The president of the republic of uzbekistan of February 7, 2017. in "strategies of action for the further development of the republic of uzbekistan"on UP-4947-decree dated 20 April 2017. in "on measures for the further development of the higher education system"pp-2909 dated July 27, 2017. "in the preparation of specialists with higher education in increasing the quality of measures for the further expansion of the network, and participation of sectors of the economy"on pp-3151-dated and this area that relate to other normative-legal acts on the implementation of tasks specified in this dissertation research at a particular level of service.

I CHAPTER. IN HIGHER EDUCATION INSTITUTIONS "BIOLOGY" THEORETICAL BASIS OF THE TEACHING OF SCIENCE

§1.1. Biology is the science of teaching content and status available

As noted in the national program for training, "the man, his identities to find the combination of comprehensive and well-being, conditions and create effective mechanisms of realization of the interests of the person, manners outdated thinking and social change carried out economic reforms in the country, the main purpose of the template and driving force. The nation's rich intellectual heritage of human values on the basis of the modern culture, economy, science, engineering and technology on the basis of the achievements of uzbekistan to prepare personnel development is an important condition of the formation of the perfect system" [2, 31-61-b.].

The president of the republic of uzbekistan of February 7, 2017. in "strategies of action for the further development of the republic of uzbekistan"on UP-4947-decree of the republic of uzbekistan in 2017-2021 years with the number on the implementation of the strategy on five priority action for the development of the state program of the republic of 367 units in secondary schools to strengthen the material and technical base, their "laboratory equipment and other stock providing with" through "activities, they get the education of students and the teacher to create favorable conditions for the use of"has been identified as a necessary task to achieve [4].

The main objective of the program is to expose the comprehension of reform and improvement of the education system. This, in turn, the responsibility of the academic literature on higher education teachers of educational sciences of the current period appropriate requirements and the latest achievements of science taking into account go to update, and to introduce innovative technologies in the educational process requires education [1, 20-29-b].

The direction of the teaching of chemistry education to the students to prepare the organization and implementation of the content of professional activities of the state educational standard (requirement and the course), training plans, curricula, textbooks, educational and methodical manuals and has found its reflection in the reference.

The level of education of state educational standards and educational downloads are necessary and sufficient to put the size of the basic requirements that defines. Education buyers should possess the knowledge, skills and qualifications will set the state standard of education. Also, the state education standard as a means of control of the educational institution will serve as the criteria which define the directions of preparation conditions. As a result of the introduction of the state educational standard of education as head of state institutions will put the task before. Proceeding from these tasks, while the institution of education, how the knowledge, skills and qualifications that should give, and how it gets defined measures to implement gradual. Policy taking into account in the preparation of teachers in the higher education system of representing the content of education given the following set of state educational standards:

1. Pedagogical conditions for the implementation of training and professional it services, which are put in relation to the general requirements.

2. The size of the training of students and their obligations.

3. Academic freedom in determining the content of professional training of pedagogical education institutions.

4. Higher professional pedagogical preparation directions (specialties) which are put into the list of general requirements.

5. And the content in relation to the level of professional preparation of graduates and development of minimum requirements for the approval procedure.

6. Be sure to follow the requirements of state educational standard of the state to take control rules.

The First President Of The Republic Of Uzbekistan, I Will.A.Karimov focus special attention to the issues of introduction of modern information and communication technologies, "at the present time the modern sector, which is first and foremost on the basis of the telecommunication network or all of these olmaylik development and information technology in general, see and experience the world state and the most advanced standing is not difficult to realize in the example," who noted that [8 - b. 280].

Prerequisite for the use of modern educational technologies in teaching the following: first, a lot of the information of the size, the width of the fact that they are updated and quickly; second, the speed of technical progress, to the fact that they quickly develop modern tools and supplies; third, has been made in the development of various technical means of teaching; the fourth, external influences, for students bundanda interesting phone, internet, social sites, etc.k.s there. Such effects at the level of modern requirements to take the lessons of today requires.

By today, the person is directed to applied technology education. All of the educational process in modern conditions reflect the real deals, according to the development of the person, and which ijtimoiylashtirish independent, critical and creative thinking abilities are required to be trained. Education refers to education that is directed at the person, which will show you these possibilities. The younger generation of interest at the present time the institution of biology and multimedia computer equipment to create animated on the subject of motivation. Thus, the development push by students on the subject that should be able to easily and quickly leads to skills and qualifications. This condition is interesting facilitate the work of the teacher, the students you want to experience at the same time you're done using the computer itself, allows you to mastering the subject [119, 94-b.].

H on various issues of the teaching of chemistry.Omonov[77; 46-b.], A. Mamajonov [67; 22-b.], M. Nishonov [74, - b.111.], O. Steshina [98; 21-b.], Sh. Begmatov [30; 17-b.], M. Ajieva [19; 120-b], E. Eshchanov [126] F. Alimova [22; - b.183.], N. Anvarov in [18; - b.155.], X Rajabov [89, 8 b.] those who work as researchers.

N.G.Rakhmatullaev, Etc.T. Omonov, Sh.M.Mirkomilov [90; - b. 320.], M. Nishonov, Sh.A. Mamajonov [73, 52-53-b.], A.Off [136]s "the teaching of chemistry" in the literature related to pedagogical and didactic education in all stages of the student's general knowledge of software requirements, imagination and skills of independent work on the basis of improving the effectiveness of scientific thinking to the interest of science education, enhancing their professional knowledge and deepen theoretical and practical sessions to increase their activity during the "teaching of chemistry" provides information about improving the structure and content of the phase. Also, this literature students of biology scienceand technology

to better understand the use of innovative information to provide information. H.T. Omonov their research in the teaching of chemistry philosophical-pedagogical principles and has found the answer to improving its solutions [77, 46-b.; 78;79].

S.A.Nizomova "Chemical education effectiveto raising innovative technologybee" in the process of teaching of chemistry in the research that develops together with students who expressed opinions about what education should give education to give [71; 69-73-b].

F.A.Alimova of new information technologies in scientific research work created on the basis of the automated system science tutorial program and using them as "inorganic chemistry" and "physical and colloid chemistry," the stage "solution and electrolytes found on dissosialanish nosebleeds" and laboratory developed a method to take classes from. The solution created on the basis of studying the didactic use of the program kompyuterli [22, - b. 183.].

Pedagogical education (pedagogy). the journal ofinteractive teaching methods at the buyers to know that they may be involved in the process of education, they know who they may be able to understand and thinking and thinking things that, in the lesson the lesson the role of the teacher bring the students to achieve their goals far partially straightening the activities of knowledge that students give good results in the control of the use of information technology provides information about [61;76-83-b.; 66; 48-51-b.; 68, 42-44-b.].

E.If the research work is paid to the complex issues of eshchanovning organic chemistry, chemical education and teaching methods in the process of the computer's introduction of modern technologies and independent schooling attention to the chemical reaction process [126].

L.T. General chemistry for students of academic lyceums in research zaylobovning oxidation-reduction method with the help of information technology to read given the coverage of improvement [55].

M.B.Ajievaning tadqistage, which is the chemical in the work of the school qotliy the use of raw materials “Galogenlar” revealed to read based on the theme of improving the methods of [19 - b. 120].

O.S.Steshinaning redoks of research in higher education–system is based on the results of his research have been studied on the basis of scientific methods [98, 21-b.]. N. N. a. “the teaching of chemistry for the use of advanced technologyfor oil and gas in various forms and methods” in the literature during the lesson, the opportunity to create efficient use of information technology, chemistry, the task of the teacher of students to enter the business is active far, to do this, teachers give their thoughts on the efficient use of information technology is reflected important results [16].

N.A.Parp, X.R. Head.G. Muftaxov “inorganic chemistryof theoretical basics” training guide in the structure of complex compounds, naming their classes the most important information about come inthe oldest and largestrilgangan [83].

Yu.T.Toshpo'latov, N.G.Rakhmatullaevof the world “inorganic chemistry” lecture on the subject of prepared compounds in complex theory to explain and chemical nature in their gardens has been noted [103].

Sh.V. Abdullaev, D.X.The muhitdinovs prepared by "the teaching of chemistry"methods from the manual of a complex combination of the texture to be taken and information on showing given [17].

V.G.Yusupov, M.T. Toshev, N.A. Parp “Koordinasion of chemical compounds” in the manuals of complex compounds of chemistrydaon place and itng how to field applied ineitherof the grass after moving tellbirths [124 - b. 298].

N.I. The observer “general chemistry” textbook written for the theme of pedagogical higher education institutions of foreign literature lit on the basis of complex compounds obtained from sources [38; 318-328-b.].

Human that is associated with the process of informatization of society, you can know the information from the internet. The use of information technology in the educational process is required from us and life itself. Young people inoil and gas information from the screen is a very strong predisposition adopters. The function of the teacher knew that the case should be easy to approach them to give you the information through [9- b. 400].

N.I.The Observer, M.I.Ibodulloevalarning for the students of the pedagogical university "Biology" educational-methodical methodical recommendations on the application of educational technology given on the science of aids [39, - b. 98].

X.M.Research in the teaching of chemistry rajabovning tarixiylik the systematic approach to the principles of chemistry and the chemistry associated with the history of the world, the universe of uzbekistan the development of the community without the improvement of scientific-methodological aspects of the problem on a conceptual basis to work out special dedicated to holistic, comprehensive research work was carried out[89,7-b.].

Academic complex compounds on the synthesis of N.A.Parp, k.f.d. professor T.A. Azizov headed by scientists conducted studies. They synthesized complex compounds in the solution of industrial and mechanical method, the content of element analysis and physical-chemical analysis was detected. Also, differential thermal analysis, which determine the structure of complex compounds with complete instruments, [86; 20, 172-b.].

J.O.Tolipova for the students of pedagogical higher education institutions "for teaching biology central to innovative technologies policy" on the basis of tests in the textbook and reproductive produktiv the application and control of the evaluation system lit [101].

A.Yu. Withdrawals are issues which will put the issues of teachers methods to take off textbooks of general requirements, the types of issues and their withdrawals, schools, academic lyceums and professional colleges of the issues in a way to take off is directed to [59, - b. 220].

According to the analysis, in the republic of chemistry teaching, modern teaching and research on various aspects of the use of information and communication technologies work] toI go, though, if exactly "Biology" flekarzem the technology of teaching based on research on the implementation of the work] toI not go. This research carried out the work butI "Biology" flekarzem the trainingof the beginning served basis.

In the countries of the commonwealth of independent states teaching and learning materials chemical computer's improving the effectiveness of his teaching and performing the experiment in the laboratory of Yu. Gavronskaya [44, 47-b.], P. S Belov. [32], O.G. Bodi Has The Features [37], E.G. Zlotnikov [57, 18-25-b.], V.S. Polosi V.G.Prokopenko [88, - b. 224]the country A.Xayitov [108, 93-95-b.], Sh.Sh.Daminov N.N. Tour, S.K. Aliyorov [48, 96-b.], O.M. Yoriev [51] M.A.Qo'chqorov [117, 283-b.; 118, 48-b.; 120; 121, 141-142-b.], R.Has Been Awarded With, Sh. Abdullayeva [105 - b.192.] Z.K. Qodirova, A.G. Muftaxov, M. Yakubova, I.Abdugafurov [116, - b. 184.], in the research work of scientists like can be introduced. They directly experience the issues of improvement of the organization and conducting of chemical technique and exercise methods to take off a life of modern methods likely to improve their contributions have added.

N.V. Kuznesova [63. 95-117-b], V.P. Bespalko [35, - b.352], O.S. Zaytsev [54, 98-144-b.], K.B. Belikov [31]s psychological – pedagogical principles, pedagogical kontsepsiyasi in their research about the structure of functions of the system has been clarified.

Psychologists's noted that noted that students read the knowledge book is very effective on the screen than take out push the development of reading used to say that. Animated, multimedia and video images, voice teach students the state of the process: to realize the essence, it will cause a drastic lifestyle. Comes in a very favorable condition for the process of education students.

Biology is the subject of the system of teaching at the laboratory reflected holda real-world modeling shows the appearance of the kompyuterlashgan educational environment. Thus, html, dhtml, and java in the physique gipermatn hm, MC flash, delphi, c and other software summary shall consist of the program group will help. With the help of this program in a certain subject subject of the content, assignments, and other issues and convenient form and in terms of the interface to be provided to the recipient of the education in the process of reading fun, visual, and ensures that the quality of engrossing.

Iboratki more of the actual work of the use of the laboratory in which to make laboratory experiences for the organization of complex devices, devices, and chemical agents of production, storage, replacement is not necessary repair works in addition to chemical wash the dishes like you don't have to get the job done. All laboratory work computer's memory or in external memory and stored in a place in the electronic option. Safe computer of the device, the user only specified is not difficult to use and require skills of working with the special programs at the same time [15].

To conduct a laboratory experiment Chemistry from the software can be used. Jodi L. Seas [53, 1517-1525-b] homechmiddle schools in california, one of many scientists in the bone marrow Chvaccine+ program (www.chemvlab.org classes are based on.Bt hejoint-stock companyribalarat also students thisrnIng thiszlaatIrIsh ko'nrsatonchI the soupton, stainednI ktratIsh mitmkn.

Slhunteniya kimyogap rtake edaoglar nataša Ran get hera [131, 38-b.], Dejanan Dinevski [49, 4-8-b.] american scientists and J. Russell [132]Shriver and atk' [115, - b.850.], R.By koz [130]lar students, thisrto knowledge berIshat an'yuanavIy usunlda va axbt oroexnlopromised thisrnI qo'llagan holda, atrslao rbecome borIshhan, t- stock companyribalarnIng natijalarI sIfatida students, thisr this knowledgerI the soupton. Bflourat an yatgrated into the landIshtonthisrInI laboranarrowiya macitycountryitlhorselarInI - typeli sabablar withn dorilmas, stained, kompьyuter texn n thelopromised to thisrI eithershlaor rasida fanlarnIng ommavIylashIshInI ta'minlacityI va an'yuanavIy usundid it nIsbatan yusnowI samaredorstained the neckinIcha understandingIrIshton.

Students thisrnIng chemical faniga qIzIqIshlarInIng analIshInI boshqa pget edaog Tuysuz, C. [104, 37-53-b], the flavors Z., A. Ayasi [99, 183-199-b.; 100, 159-170-b.] Ishlarida also ktratIsh mitmkn. Chemistry atrslarInI obecome borIshat Macromedia Flash atst payida tayeitherrlanhan kompьyuter animasiyaladiridan keng foy'sfieldnIshI natgratedsida students thisrnIng chemical fanInI thiszlaatIrIsh ko'nrsatonhInIng analhan, stainednI ktratton.

Jahon literaturetlarida axbt oroexnlopromised thisrInIng thetI sIfat, ida talaba thisrnIng qIzIqIshlarI va chemical faniga nIsbataq narashlarInIng thisztorIshI known Izohlamnton. Bflourathu ylosagain comeshnIng asaucely sabablaridan brI sIfatida thisback atrayonInIng mapureavIo ybecome borilishI known Izohlamnton. Literature, internet resources andIatn keraair of the laboratory for informationlarnI olish mitmkn.

Currently the laboratory of inorganic chemistrynI completed the practical work on the creation of a number of foreign countries. For example, the virtual Chemistry laboratory carnegre mellon university in the usa made by molding it from the internet or a disk can be used. Experiment two-dimensional graphic models is carried out using. Chemical Lab at brigham young university in usa Virtula cd-rom of three-dimensional graphics offers of jamlanma virtual laboratories for conducting a chemical experiment. Oxford university (uk)solve problems videolavha do experience it offers. (Virtual Chemistry-http:/www.chem.ox.as.uk/vrchemistry).

The delivery of the subject to students in the educational process is important because it can distribute on the internet or in the drive. All students do not have the same opportunity, so they will give good results from the use of the library for help. Another problem, that is, how should you use it from the teacher during the lesson that can be solved on the computer. Training can be a different approach in the organization. Topics can be delivered through the internet or compact disk. For example, the complex compounds on laboratory experience using animated bumps can learn. Thus, the biology of science information technology in educationIget from the use of the knowledge of the students increases the interest you will develop it. The biology of the essence and importance of science leads him to realize quick and easy. Safe implementation is to carry out chemical reactions.

Natural (chemistry) the teaching of other disciplines such as science education unit, according to the younger generation, political, ideological, moral and physical qualities, and their development will set a goal to mind the form of high culture.

Therefore, the young generation, the future of the specialist, in general, how the nation's political, ideological, moral and ideological principles of continuous education on the basis of performed work and live in our country today is directly dependent on the efficiency of the system.

The implementation of such important tasks in natural sciences (chemistry) recent advances in the field of scientific innovation and to rely on in the case, the position available scientific-theoretical study, content enrichment, improvement and reform to meet the demands to the extent that it is necessary to bring the principles to focus on. Check out the quality of the success of the system of science, reading, mutual integration between the subjects of "continuing education objects and subjects" the students in the fullness of the mind and the effective organization of the educational process aimed at the development of thought related.

That came out, enrich the outlook of the youth of today in their devotion to the motherland, to its development affected the sense of national pride formation, found them growing in the spirit of national and universal values, explain the current times has dramatically intellectual-spiritual nurturing highly qualified professionals who is able to make independent decisions that can respond to competition in the higher education system taught not only of humanitarian and socio-economic sciences, as well as, remains one of the important tasks of standing in front of natural sciences.

Educational objectives the program educational objectives and their reflection in teaching the subject, although it has found important, yet the teacher science news: scientists describe the students must go to the country that bring this area of research. The goals and objectives of the lesson this in turn not only educational, but also is considered important to achieving its educational goals. Students in this thing, national pride, loyalty to the country, that increases their interest in the profession [28].

Changes occurred in the world in the field of education and spiritual enlightenment of the age of the globalization process of thinking and outlook
that plays an important role in formation, who enrich the spiritual world, the spiritual emptiness that does not allow to arise in their minds, humanitarian and social–economic sciences with the natural sciences will require improvements in the levels of the content.

Consistent interaction for the teaching of science competition and to ensure the improvement of textbooks and manuals, they reflect updates in the field of science, present, and future lighting tasks, independent of education, teaching in the process of extensive use of advanced information and communication technologies to take into account the capabilities of modern plays an important role.

The creation of electronic educational sciences educational tools science teaching in further expands the possibility of the use of modern information and communication technologies. This, in turn, is the main factor of mastering deep knowledge of students on this subject, if the quality and efficiency of education increases.

Men embodies the positive qualities of high moral character, that is, the above-mentioned words are a highly skilled practice, ongoing reform of the education system are launched in the country on a progressive basis and hence, increasing confidence come to justify the quality of education. Therefore this innovation development - a factor of sustainable development.

The cultivation period of competitive personnel have become a pressing issue. The experience confirms the viability and efficiency of the market economy of the world. Attentive aspects, that is, the production category, the concept of the need for production personnel, as well as the quality of their preparation compared to standard defining the requirements for the main customer of the process in terms of providing financial and material and technical personnel system as participants appears.

To be highly qualified staff of creativity and innovation is realized due to the increase in activity only. Their ideas, to carry out the scientific directions, which in the area, no matter if the product or service bozorbop out of showing their commercial work move, and in this process, young professionals, attract students, and can achieve efficient and effective outcomes. This in turn educational process, scientific and intellectual labor as the chances of their life through the integration of the production environment to meet your needs. So until you can the study of mental labor, financing, development aylanmokda order to request a transition to the times.

The main purpose of education: the level of specialization and cooperation between having competitive in the labor market, responsible, a master of their profession, which is freely doing yondosh occupations, who could do an effective job on their specialty at the level of world standards, and regularly consists of the preparation of qualified professionals who work on their own. These ready lessons in the process of direct qualified personnel and their knowledge, skills and qualifications of the following o'zlashtirishganlariga depends on.

Until that time, only prepared in traditional education students possess the knowledge to teach, had come. Such methods students in independent thinking, creative research initiative was because of an interaction. Teachers - scientists for years in the education system “why are we teachers? What we teachers? How teachers we?” in addition to looking for the answer to the question “how teaching can be an efficient and effective?” - looked for the answer to that question. This, while scientists and practitioners to texnologiyalashtirish of the educational process, that is, to read a technological process that gives guaranteed results related to the production of you can see clear of the idea that led to trying to become.

Many researchers in different countries of the world in education innovasiyalar he always conducted research on the application. They are “innovative”, “interactive methods”, is a collection of good information on innovative technologies. In recent years, independent innovation and scientific and methodical research on the implementation of information technology education on the basis of implementation of education and experience shows that can solve this problem [58, - b. 256; 93, 113-b.]. For example, innovative technologies "brainstorming", "Pinboard", "Cluster", "Design", "Assesment", "Case-stadi" technology independent learning of students on the basis of the implementation of.

Apply today innovasiyalar to carry out the following tasks in the education process requires:

- determine the exact purpose of science education;
- determine the content and size of the fan;
- recommended and required the development of technology to education;

- the creation of material and technical supply of science;
- study the characteristics of the recipients of education.
- the teacherng lessons and training design.

Xorij educationIinnovasiyalar at the following innovative pedagogical technologies on the application to dwell on it should be relating to the information that we know.

Modular learning technologies, teaching is one of the promising systems because the recipient of the knowledge of education to get creative abilities and opportunities are best adapted to the development of the system. The purpose of training is expressed through the activities of teachers in traditional education, the knowledge that is directed to give the business buyers were expressed by module teaching and learning, professional activity will be directed.

Modular comprehensive learning of modern education creates the possibility of withdrawals of the following issues:

Module – faoliyatli we have teaching on the basis of content and optimization of the program tizimlash a variation, adaptation allows achieving greater interoperability;

– individuallashtirish to read;

– track action at the level of practical activities to teach that the teaching and assessment of the effectiveness of monitoring;

– to enable the full realization of professional interest and learning opportunities on the basis of independence.

The effectiveness of modular training depends on the following factors:

– educational institutions of the material-technical baseto;

– the level of qualified teaching staffto;

– the level of preparation of students;

– kutiladigan to evaluate results;

– the production of didactic materials;

– the module result and analysis.

A shortened through modular educational programs for teaching tabaqalash full and thorough step-by-step to create an opportunity to read so that you can be individuallashtirish to read.

In the transition to a reading module is pursuing the following goals:

ensure that teaching is continuous;

individuallashtirish to read;

to create conditions for independent mastering of educational materials;

to accelerate reading;

the effective adopters reach of science.

Modular summarize the main issues of the science of teaching lectures, allowing you to read with your guide and problematic oriented data that requires it. The development of creative abilities of students, lectures yo'naldiresurrectionmog'i should be.

Lectures and practical laboratory sessions, along with the structure of the module, they should be supplemented with lectures o'rganiladigan new content material.

Modular credit system. High scientific and technical progress in the second half of the twentieth century was to reach his peak. Fan - technicala development of the (ft) shundayki the current level, the development of technical sciences and technologies may be developed at the country level rates achieved in a particular country which does not ensure it is taken. Ft scientists and specialists of various countries of the cooperation of not only the development of the next scientific-research as a result of the integration of the work can be done. The level of development of science, engineering and technology is closely related to the development of education, we can consider that, to the rule of integration in the field of international education is to become a clear problem.

This technology creditng be the creation and use of topical determines that, because of international integrallashuv in the field of education, primarily in the educational processng will rely on technology to establish credit based on.

Modern technology, teaching, higher education muassasa the interior of the teaching staff only scientific level and unvonli, to be composed of highly qualified personnel, high-quality teaching is required for the initial requirements. This system has the following specific properties of the organization of educational process.

All study subjects are divided into two groups - binding and, on the basis of student's choice subjects. 1 about this:it is taken equal to 2;

individual study plan for each student will be able to.

Main document transcript (transsprint rourds of it's features unifikasiyalashgan document is a single shape is formed in teaching to recognize the results of a mandatory document. Transkriptda, the student's state assessment system adopted in also taking on the system featured in the price of credit taken is featured contains information on:

the number of subjects in each semester o'rganiladigan Z-5 units;

work independently allocated to the hours a lot from the audience.

Credit for the master of science in educational represents students at all necessary costs of labor. This isthe number of educational crediting of science, in most cases is equal to 3.

Teaching the application of science content (sIllab the) following their include:

- its full of educational sciences curriculum and the order number on the naming;
- the purpose of the study of educational sciences;
- a brief of the content of educational science;
- calendar plan, training schedule;
- teaching technology;
- the responsibility of the student and the demands put upon him;
- student's knowledge assessment procedure and criteria;
- a list of basic and additional literature.

Educational downloads in the calculation of the teacher, it isthe amount of time spent on a work of wax (the audience and independent work hours) in the organization of the academic year 920-950 hours - will be taken into account.

*This*qitish the basis of the process:

- directed to the person teaching technologies;

- student's independent work, the teacher participates as a consultant;
- teachers and students of academic subjects has the right to choose.

Distance education, foreign countries, teaching the theory and practice of the rich experiences ofI, the direction of researchaccounted for its relevance confirms. New pedagogical information and telecommunication technologiesIfrom the use of education based on one method of distance education draws us to understand the essence of which is.

Gumanistik exactly which form of integrated synthetic and distance learning education education reform education system in the conditions of theoretical and practical issues facing a big part of the problem should be solved. Distance learning education in foreign countries and the establishment of educational institutions studying in distance education in the process of development applied to the organization of various forms of distance and the means of providing education to the analysis of didactic and options, it is necessary to focus the attention.

Distance learning - distance learning and information and educational environment as a means of sharing information using a basis for is from the collection of special educational services. Distance education means education information and education environment designed to meet the needs of users of information resources of the mutual relations of the protocol data transmission hardware - dasturli and organizational - methodical and systematic organizational from the collection. Distance education - teachers teach students to: basic to deliver the volume of material in the process of training students of the mutual interactive communication of teachers and students teach students to independent reading material to allow you to work independently and on mastering skills and qualifications and the assessment of the knowledge they received in the process of offering information technology pack.

Web method. This method seminar or conference in the lesson via the internet, students are now at the same time with audio, video (and many other interactive opportunities listed in the previous post) and go get live with the next use of for writing this lesson can be taken whole or classes within a single platform is an integral stage in the educational process of mutual link that can be said to be a disposable lessons separately.

Problematic methods. The essence of the method is problematic in the training process on the basis of creating a problematic situation and offer to take off his didactic in the conflict lies. Education kontsepsiyasining the main problematic concept of "problematic situation" "the problem", "find the problem" is like. This method is problematic situation at the initial appearance, it represents new knowledge to eliminate the problem of exact or partial understanding of the subjects, methods, and requires the will of including the development of movement skills. If a student is looking for ways to lose the primary to find data challenges in solving problematic situations obviously won't accept it, that the solution of the problem are not reflected in his mind.

Be expressed as the problem of thinking understood the essence of available knowledge, skills and research skills, and complex problematic on the basis of experience begins to accept the situation as soon as possible. To help you find solutions to problems in the process from the use of this method is shown and is not limited to giving directions. Seem to show how the parameters of the solution is in the problem, it is a problematic issue.

Man always will address problematic issues. It will become the problem in front of you appear to be a problematic issue, that is, a system of knowledge known to rely on itself for its solution and many times show gets set for your country. Failed in the case and search for other issues in the indicators loyihalaydi a new option on the same problem.

An intuitive teaching method. Application of the method to the students in collaboration with teachers in an intuitive reading will identify issues that needed to be solved. Students mastered the necessary skills independently while they take the issues in the process of the proposed research, and its solutions was compared with other situations. Withdrawals installed mastered the skills to know students during the scientific methods to carry out activities tadqiqotchilik issue, they take it.

Check out the main application areas of the educational process and the electronic textbook in the different sessions, the main subject of study by students, which is difficult to master the modeling object. The volume of information is large educational complex if the budget is less than students in real time, in which case the application may be considered in accordance with the purpose of the electronic textbook.

The training is likely to invade computersьyüter technologies by means of the organization to the requirements of the time to create electronic textbooks is becoming. Electronic textbooks see various information-text, tables, graphics, sound, animated, videotasvir, performs the functions of using music collection and storage. The electronic textbook "human computer" for the improvement of the new line of interactive communication, it gets wide and comprehensive information of users, active in the educational process, will increase the effectiveness of education, ensures a long time in keeping the memory of the subject, independent increases the opportunity to get education, computer literacy of students has increased.

§1.2. Biology" general features of the content and teaching of science

"Biology" improving the use of innovative technologies in science teaching, quality of teaching approaches, and theoretical study of the research problem in our country and abroad from the analysis of the pedagogical aspects of information about are made.

Aimed at addressing the problems of informatization of the education system from the years of independence of measures are being developed. In particular, the republic of uzbekistan, the cabinet of ministers of the "national information network and create a network on the use of filtered from the world is"on, "the year of 19 the president of the republic of uzbekistan "on measures for further improvement of the sector, information technologies and kommunikasiyalari on" UP-5349-the number of times the time takento it may be noted [3].

Of the republic of uzbekistan "national training program"of accumulated experience on education reform at the analysis, synthesis on the basis of the country's socio-economic development of the prospects of upgrading the system of personnel training in accordance with, the resources of the institutions of education, personnel and further strengthening the information base, the process of education, continuous education global information system computer information that are connected to the network with a network urgent tasks related to full coverage given [1, 29-b.].

The constitution of the republic of uzbekistan, the republic of uzbekistan “on education”law, “national training program”, I will first president of the republic of uzbekistan.A. Karimov and the president of the republic of uzbekistan Sh.Assign on enhancing the effectiveness of education in the works of ideas on the formation of student identity qadriyatli approach, which represents the content of the activity as a creature of the person's physiological and social visions and theory to know, as well as the dynamic development of the person and the social philosophical approach to the mind in the area of psychological, pedagogical consists of qonuniyat [2,31-61-b.; 12 - b. 488].

The president of the republic of uzbekistan of February 7, 2017. in "strategies of action for the further development of the republic of uzbekistan"on UP-4947-the number of times, April 20, 2017. in "on measures for the further development of the higher education system"pp-2909-number, July 27, 2017. "in the preparation of specialists with higher education in increasing the quality of measures for the further expansion of the network, and participation of sectors of the economy"on pp-3151-resolution and this area that relate to other normative-legal acts in the present day is one of the important and pressing issues of the program.

In the higher education system "national training program"stemming from the task of scientific, technical and technological development of the future teacher with today's level of kompetentlik inconsistency arising between the process of improving the quality of a number of professional identify, and implementation of modern information technology in overcoming them is not enough available that showed the relevance of the problem. The identified problems of theoretical and practical base, as well as the preparation of teachers in higher education institutions, the introduction of modern information and communication technologies in the process of promising directions are set, only the information of the higher educational institutions of the country-through a single portal and e education-information-educational environment to create a modern information technology as a factor in the introduction of the bachelor of professional reference to improve preparedness, e-multimedia educational-methodical complex of pedagogical create and create the conditions necessary for a boost axborotlashtirishni her toned introduced to the learning process [60, 79, 80, 69-71-b.].

Therefore, future teachers of chemistry education in the higher education system to pave the way about the process of preparation for re-are required to be considered. These future teachers ready in the system of higher education, especially secondary schools, vocational colleges, students of pedagogical higher education institutions of modern technology teachers who teach the process of preparation required to the specific conditions, together with intersubject communications in the case of the series that rely on the principles of consistency, the organization of teaching and educational process makes it possible to require.

Institutions of higher educationmany years ago the students personal development – their social-professional personal interest direction, independent thinking, creativity, and the active line, the enhancement of the relationship be enriched, behaviortolerates turg'unlashuvi and outlook, self-nurturing is characterized by the formation of the needs to be. The process to get education for future professionals in higher education institutions – this man in the professional activities of the successful implementation of the professional roleimportant to having qualityproducts, knowledge, skills and qualifications in the development and self-improvement for the most acceptable period. In this process the students knowledge and to master to create their logical structure, the organization of psychological them in professional activities in the run - pedagogik activities with the tie and the effective application to occupy ways, values and motivationto option their mohiyatanby emergency will go [21; 52; 65, 48-b.].

In 60-70 years of the twentieth century the use of the framework of information and computer technology appear in the theory of interactive learning and distance education. Institute of information technologies in education (gusto ii) Vladimir Kinelev, commercial pit, Boris Kosiklardan founded in the opinion of the research team, from the computer that started with the emergence of the idea of using them as a teaching tool in the 1970s, an electronic programs appeared first acting teacher. At the beginning of 1980, while using the computer begins to develop more complex software system in education. The teaching process used in the development of the internet network in the middle of 1990, began to find their reflection in education.

O'rganiladigan all of speciality subjects in higher education institutions only providing information, teach did not remain confined to the function, but the idea that develops integrallashtiruvchi, the function also has to be trained calamities. See also every science teacher to the students not only the subject that is available in the structure of scientific armed with information, but also provide them with the development of scientific concepts and other concepts who they occupy in the spirit of national values and integrallashuvini put the task of bringing up too. Therefore, each teacher ko'rishida regular training sessions to prepare learning materials to comply with the principles of the above-listed requirements and, in case of appropriate understanding to be selected and it is presented to the students in accordance with the purpose expressed in a productive use of [110 - b.256].

Teaching biology for science teachers for the implementation of this order it is necessary to perform the following tasks:

Biology education students in the process of spiritual and moral upbringing and educational introduction to the practice and development of effective forms and methods of work;

The science of biology the process of teaching and innovation in the use of information and communication technologies;

Speeding up the process of learning of science in biology from using the system module;

science chemistry education students in the process of development of science, the uniqueness of the content, proceeding from the area, forming private kompetentsiyalarni appropriate;

the didactic and methodological supply of chemical education of the new generation of the educational process, creation of manuals and introduction to the practice.

Relevant to biology education, as well as educational methodical manuals and didactic introduction to the practice and development of complex software of the new generation;

On the basis of this order there are modern approaches to biology teaching science in higher education institutions of the following:

1. Is a systematic approach to the process of teaching of science in biology.

2. A modular approach to teaching of science in biology is the process of.

3. Innovative approach to the process of teaching of science in biology

4. The person-oriented approach to student in the learning process of science in biology.

Actual problems of the teaching of the biological sciences to identify specific features of this approach are listed for each lesson and are considered separate.

“Biology,” science for the teaching of modern education technology for the use of the necessary conditions are the following ones: first, a lot of the information of the size, the width of the fact that they are updated and quickly;

secondly, the speed of technical development, the fact that the rapid development of modern tools and hardware made;

third, it has been made in the development of various technical means of teaching;

fourth, external influences, for students bundanda interesting phone, internet, social sites, and x.k.s the availability of such effects at the level of modern requirements to take the lessons of today requires.

By today, the person is directed to applied technology education. All of the educational process in modern conditions reflect the real deals, according to the development of the person, and then ijtimoiylashtirish independent, critical and creative thinking abilities are required to be trained. Education refers to education that is directed at the person, which will show you these possibilities.

This education cooperation and the relationship between teachers and students freedom of choice, based on values creates the conditions for students to be viewed as. Usually, the person is directed to the difference of the following types of educational technology:

1. Modular technology education.
2. Problematic technology education.
3. Interactive educational technologies.
4. Individual education technologies.
5. Distance education technologies.
6. Computer technology education.
7. Education cooperation technologies.
8. The project technology education.
9. Software technology education.
10. Differentiated learning technologies.
11. Developing training technologies.
12. Game technologies.
13. Gender education technologies.
14. Power-saving technology education.

Used in biology classes to the person oriented education – the student's thinking and action strategy, given his personality, the specific features of the development of education, aimed at the ability to.

Cooperation in education in the educational process of students, in teams, in pairs, in small groups and mastering knowledge together, mutual development is supported and developed, "the teacher-student(s)" the cooperation of the organization of relationship education that is representing, the idea of his basic training assignments in the team, in pairs or in small groups together, mutual co-operation is considered complete. Biology lessons, co-teaching is considered important. Moreover, in the team of the laboratory in small groups or in pairs together, increase the ability of independent thinking along with them in mutual cooperation team to lead students complete their work due to the failure of a mutual friendship and respect to each other is the form of the relationship.

Education areas of cooperation:

-to establish a relationship on educational cooperation;

-individual approach to students based on the idea of humanitarian;

-professional spiritual unity and the decision of the educational process and will find to achieve.

Educational technology in the educational process of students of cooperation in teams, in pairs, in small groups and mastering knowledge together, mutual development is supported and developed, as well as "the teacher-student(s)" of the relationship tavsifdagi cooperation organisation offering educational technologies.

The structural elements of the technology of education of cooperation:

1. Base outlines (chemist, mathematical and chemical formulas, thesis, comment on pictures, brief conclusions, symbolic characters, schemes, graphs, tables, diagrams)from the use of.

2. A test based on the knowledge test.

3. Mastering assess indicators of students.

Technology education on the principles of cooperation:

- mutual pair and small-group members of the unit;

- the responsibility for each member's personal and group success in pairs and small groups;

- cooperation in small groups-based know to organize educational activities;

- the overall rating of team work and group distribution.

The symptoms of cooperation of educational technology:

- students are the person to pay attention to individuallik;

- ready knowledge mastering and re-work them out to deny;

students ' development of independent and critical thinking;

- ensure the occurrence of a positive attitude toward teachers and peers;
- students in the development of cultural communication skills;
- create an environment based on equality and mutual cooperation.

They receive training in the use of educational technology students in the following forms of cooperation:

1) in the team;
2) in a small group;
3) in pairs.

1-option. Get education in the community. Each student's team because the team will be aware of the achievements of the successful development of learning materials are interested in by each student. After all, the success of each student and his achievements of the team to put the team before the solution of the problem depends on. Education team from biology, especially when used for the development of independent education is very important. Any chemical or students to be part of the technological process and the team will be distributed. The team together in a holistic project forms part of the same process.

2-option. Get training in a small group. Combine small groups of 4-5 students, students, and educational materials given in case of works on the block. Each student will work on the particular material. Students in a group and the other groups from 1 to take students then learned information to their peers after being given himself as an expert on the material group to which one, introduces its proficient team members with information from peers. This partnership will listen carefully, note the necessary information is required to go to his notebook. For example, the group is a small group a few in the biological sciences, and the synthesis of a particular variety is required to give to them.

At the end of training, the teacher will ask you to answer the questions on the topic studied from any student.

3-option. Work in pairs. Will be to pair students up. Each pair gets a special assignment on the topic holistically. Members of the pair working together on the basis of the development of a given material to themselves. Then they give the information about it in front of the team.

Biologiyada applied technology used in education in collaboration from students, the following is required:

- achieve cooperation with the partner and guruhdoshlari;
- active work, the question in relation to a responsible approach;
- positive partner or being in a relationship guruhdoshlariga;
- not only his achievements, but also the partner, and to feel the responsibility for the success of the group;
- work in pairs and in groups – to make you feel serious and responsible work.

Cooperation in technology education "teacher-student cooperation" based on the principle of the following forms is used:

small carrying out of research works;

- to participate in competitions, which are founded on science;
- olympic calamities science preparation;
- preparation of joint projects;
- creative collaboration in the publication of scientific articles;
- create educational resources in creative collaboration.

The use of educational technology cooperation procedure:

- students (4-5 people) will work in small groups;
- only for the team is provided with the training materials;
- each group will find the answer to a particular question;
- expert group shall be formed;
- this group is more acquainted with the work of each group;
- able to assess each student individually, the activity of the expert group;
- collect points to summarize the activities of a small group of students that joined assessed;
- the team who got the highest points is considered the winner.

"Ajurli saw" (fr. "ajour" – let alone a second, let alone past two open side) in the case of a holistic strategy, the topic is divided into several parts, lighting the contents of searching on the basis of students mastering them, others disabled to be able to provide their knowledge to the delivery. In this process the students they work with the prepared text.

Tavsifdagi innovative control tools. In modern conditions, not only in the organization of the educational process at the same time, also in the control of the educational activity of the students used to have recourse to the description of the innovation. None of the two in the same place of such a tool – educational portfolio and “Assesment” dwell on technology.

In recent years, teachers in modern education professional, while the improvement of the content of the student's educational activity, as well as on the basis of certain indicators of the quality of both the type of activities enough, unbiased assessment in relation to the increasing need this, especially, in the context of education has independent significance. Portfolio evaluation to the activities of teachers and students, to what extent the possibility of an indirect analysis that serves to provide quality and effective. Finally, the teachers and the students of the portfolio of them to the person indirectly in the examination process with different activities gives you the opportunity to introduce the members of the commission.

Technology assesment of students knowledge, skills and qualifications to the level of comprehensive, unbiased reviews assignments package offering the possibility of it biografik survey, a description of accomplishments in the field of education, the students individual assignments, group discussions, interviews, creative work, tests, an individual who abides case study, presentation, expert tracking, and business was founded as a role from the games.

Technology, basically, the following will serve three purposes:

1) the students knowledge, skills and qualifications of a comprehensive, unbiased assessment

2) students of knowledge, skills, identify opportunities and skills development;

3) knowledge of students, the development of skills and qualifications which perspective plan (programs)see the formation.

The didactic significance of the project for technology education. Project – specific plans, on the basis of natijalanishini cooked without guarantee its target, the content of pedagogical activity aimed at trying to work out significant results. The program, project, model, and other technological map appears on the form. The scientific basis of the project description or the idea of having a creative pose.

Design – based on the initial information, kutiladigan to predict the result of prophecy, or the process to produce the content through planning activities is focused on practical action.

Teachers have to create the project:

- create a project;
- process stage by stage lighting;
- set a clear goal;
- suited to identify the target task;
- educational content material forming;

questions and assignments of the system of development;

- methodological base structure or process of the event;
- student evaluation of the level of knowledge of the diagnosis and its level should be able to tarbiyalanganlik such as skills and qualifications.

The design of the educational process – effective for the organization given all the factors obtained in the special education process, develop his project.

Qonuniyat of the design of the educational process:

1) the effectiveness of the design all of the components of the educational process (manufacturing process, manufacturing process management, tools, information, socio-economic supply)shall be provided with illumination in the project, in accordance with the purpose;

2) the training depends on the individual characteristics of students with and without technological tools is selected;

3) the design strategies of the methods is selected according to individual teachers;

4) design of feedback quality (between teachers and students) scale, design, content and all factors depend on the efficiency of.

Teaching activities during teachers in the education process of their work, or uses one of these methods. “Methods” in the word “what?”, “How do you need to?” is determined by finding the answer to that question. Complete certain tasks in the sense of understanding the techniques, the means of achieving specific achievement goals, methods and ways to understand. I can say that this concept came out from the teaching methods of teachers and students, teachers and the students ' work and its method is determined as the means of knowledge, qualifications and skills is achieved by adopters, occurs, and outlook ability develops. While all future teachers of social knowledge, qualifications and skills on the subject of mastering in umumixtisoslik plays an important role. Social sciences practical as a preparation on the basis of professional preparation of the laboratory is allocated in the learning stages takes shape.

Laboratory work, demonstrations and experience with independent work. This type of training is of great importance in the process of teaching chemistry. Biology is an experimental science based on relevant theoretical knowledge on practical experience term to conduct an experiment to demonstrate to him since, laboratory performance issues and withdrawals can be mastering of the work on searching. Therefore the science of biologyteaching students to theoretical knowledge together with this knowledge to strengthen, deepen, expand, and laboratory work practice with tie for a show experience, independent education is one of the important components of the organization and conduct of the educational process.

In the process of research will determine the presence of some problems in practical teaching of chemistry. In particular, in some cases higher educational institutions of chemistry in teaching and laboratory equipment all lacks. In such conditions it is necessary to conduct in the form of training and the virtual organization. This particular shows that should be conducting research work.

Control works. Higher education institutions in the biology science, control works, check out the results from the theoretical and practical aspects of students mastering the educational material associated with it.

Independent work. Independent work of students in individual, or group can be juftli. Mutual help among the students independent work in the process group to the display of the available experience from the positive side of their organization is known.

Thus, the chemical potential of the preparation of teachers of technological education within the specialization of deep technological knowledge, you can look at the result that the development of qualifications and skills. The result of the activity of the teacher – students at the level of possessing knowledge of looks. These are: simple product of the activities on the preparation of technical and technological knowledge, the results occupational adaptation, technological please complete the operation on qualifications and skills, and other processes is manifested in the organization of independent work.

Teachers training manuals and audiovisual ko'rsatmalilik principle of diverse teaching (hear-see) the use of technical means, in terms of pedagogy is reflected in their made in a reasonable manner [115].

Ko'rsatmali manuals — students o'rganiladigan objects, events, processes, pronounced upon (mainly) see used for the purpose of training in methods to ensure cross means. Master a description of the knowledge, which is available in the student imagination, understanding, life and work experience, depending on the task of teaching aids of the lesson konkret ko'rsatmali the differentfeaturesI performs. They as a source of knowledge, as well as to give teachers delivered, explanation, conversation can serve as the picture at the time of use. Often, this can lead both tasks in a complex way.

§1.3. "Biology" sciencefor teaching modern texnologiyathe worldfrom the use of the lesson the effectiveness of oshirish factor

From the second half of the twentieth century, from the era of the transition period in the industry of information technology in the sphere of education as a result of the need to strengthen the hold information. This in turn each chemistry knowledge within the topic specified in the curriculum and the teacher did not remain confined to the master program, but creative maintain the functioning, the research work to be done, the development of creative and critical thinking has noted that one of the important tasks is to create the conditions.

The introduction of innovations in the higher educational institutions of our country to teachers in the educational process plays a key role in each. How yangilanmasin the content of education, new programs, standards do not create to be at the level to which the process of implementation of their education is, above all, related to the teacher remains. Therefore, we studied the chemistry teacher in the teaching process information daily on a continuous basis, the full implementation of the results of his work time to be at the level of the requirements of the educational process is achieved.

2001 the cabinet of ministers of the republic of uzbekistan of
August 16 in "on the approval of state educational standard of higher education"on institutions of higher education with the decision of the state education standard, umumixtisoslik and social sciences from educational-methodical complex of a new generation who put the production of the general requirements approved[1]. The adoption of the present resolution plays an important role in the development of the higher education system. This state educational standard, according to science in higher education institutions, in particular, kompetentsiyaviy approach is based on the teaching of science in biology. Also, this decision was also conducted in the same area on its own independent research work of teachers and regular performance of the

---

[1] 2001 йил 16 августдаги "Олий таълимининг Давлат таълим стандартларини тасдиқлаш тўғрисида"ги 343-сонли қарори.

educational process and develop a mechanism like the presence of defects that form yo'naltiruvchanliklarni small effect was noted.

Multimedia technology is considered important in today's modern education system and the impact of its important features to take the interactive information to the user in the work environment is capable of. Speech and multimedia computer speakers while proektor appear to be of sound, video and animated in collaboration showcases all the necessary aspects as quality of the organization who gave me the chance to demonstrate and prepare a visual presentation materials.

Multimedia technology - allows you to use one of the methods at the same time, few information available: text, graphics, animated, videotasvir, sound and so on.k.

During recent years, many multimedia software products has created and is being created: entsiklopediya, the tutorial program, and others have also reported the presentation computer.

Multimedia products:

First, the user, of course, that provides interactive interaction, that is, the exchange between human and computer bo'yruq and answers can ensure the environment dialog that creates software;

Secondly, various video effects and ayguo used in the environment.

Macromedia director – serves to create a presentation and multimedia products. This program MMX-technology, as well as, buttons, slides, clips and allows you to work with graphics.

Multimedia applications are the following:

presentations;

animasion rollers;

games;

videoilovalar;

multimedia galleries;

audioilovalar.

Basic principles of the form of ict teachers and professional orientation of the learning process the complex information to consider as the new year you can. The principle of professional orientation of teachers in ict personal interests and professional activities in accordance with the active use for the preparation of educational objectives and the subject of the relations of the united improvements theoretical education is provided for. In principle, the required level of the education system complex information in pedagogical formation have reached the stage of ict to enable the provisions of the ec noted that the complex is provided for the character.

The person directed to the introduction of ict in the education process in the context of education assists in increasing the quality of education. Teachers from using ict education of the person directed to the formation of effective prepared to organize the process of education that leads to an increase in [29 - b.232].

Today, information-education system, computing and communications equipment, software and staff to be complex, which includes the process of education to meet the information needs of all participants in the complexity of the education system have provided information model to support dynamic change. Information system oriented to person's education education education in the context of the whole application and information-education system has been about.

In recent years, the focus of scientific research work computer, a local network, the global internet network as well as psychological-pedagogical education opportunities (in all its stages) is being put on the issues of the use of options.

Modern information technology (AT) psychological-pedagogical capabilities studied in exactly the same context, the current process of their education, you will be able to work out effective option. Personal and pocket pcs, network resources bilish and communication expands.

The selection of educational resources – existing conditions came from in the case of teachers, students tanlanilishini the organization of the training and education of the means proper for them on the use in the educational process is provided. Reading strategies to read and well – established ways to convey to the minds of the students of educational materials, that is, the chosen teaching method, for use in the practice of the form will be provided.

Education tools

1) textbooks;

2) manuals;

3) tables;

4) newsletters and didactic materials;

5) the way of referrals;

6) technological maps;

7) a map of the typical errors;

8) includes assignment sheets.

The choice of the type of training – the skill of the teacher, according to the didactic purpose of the training, the content of educational material, the time allotted for training, as well as the necessary conditions tanlanilishini represents. The technological map of the training – training make up the main core of developments. It varied depending on the type and method of training can be chosen.

In the educational process the following types of models are used:

1. Educational models (used in the process of education; ko'rsatmali weapons, visual tools, educational programs).

2. Experience models (scientific, practical experience in the conduct of is used; the scale of the object loyihalashtirilayotgan kichiklashtirilgan or copy).

3. Scientific-technical model (used in the research of processes and phenomena; device, device, devices, equipment and machinery).

4. The models of the game (carry through various actions in different situations, objects, skills, applied skills in order to ensure the computer, sports, economic, constructivism, games, etc.).

5. Imitasion models (real or simply to accurately reflect realities at this level, but not exactly like him in order to be used; practical actions which serve to fulfill the various simulators, mechanisms).

The effective use of interactive methods in the training process and holistic, total process management chemistry education from the teacher to the organizers to have the ability to show the particular activity, topqir, requires a determination. At the same time the speed of a particular action to be able to stand in a certain position should be equally affected and the students know.

Chemistry teacher training sessions and study the effectiveness of using interactive methods the correct application of the applicant should pay attention to the analysis of the results they obtained. Qobiliyatinigina students in interactive methods using independent thinking, but it is not certain to come to a decision on certain issues of the disabled in accordance with the purpose of nurturing both.

Chemistry teacher as caregiver should help to increase the educational activity of each student. See this help can be in its place the following:

- students ask questions;
- students answer the various questions which come by the appropriate return;
- a conflict arising between the students, unclear problematic add clarity to the debate;
- assist students on certain issues to come to a single conclusion;
- according to the needs of students, the necessary information emergency stand ready to give out information;
- create the opportunity for the group to decide on certain issues.

The human factor in modern education as well as humanitarian of the idea should ensure the rule. For this reason, the chemistry teacher, not their interests, on the contrary, the interests of students, respect for their rights, pay attention to the pedagogical conditions necessary for the creation of their business diploma, will be required to seek opportunities in this way.

Teachers teach students in chemistry: science and the direction proceeding feature, interactive techniques for use in the selection of a particular lesson, paid particular attention to the form of qualifications. For this purpose, the necessary theoretical and practical knowledge was given.

Of education in the selection of the method according to the purpose to apply the following general conditions:

1. In higher education, the teaching of science in biology in accordance with principles come to achieve.

2. The science of biology theme feature o'zlashtirilayotgan get, considering the content.

3. To achieve the goal of education and teaching to provide the services make sure to take positive solutions of the task.

4. The educational opportunities of students (education level, specific aspects of the team and others)to keep attention.

5. Laboratory work and experience of the subject, as well as, with practice, take into account that is associated with that life.

6. The personal options of the teacher (experience, level of training, educational training situations which arise in the process)chamalab on to know.

The effects of various strengths in the education process of random practice (unknown) arising from the analysis of the situation than in advance of you can't get their effects. It is necessary to be prepared for such a situation, the teacher. Given the situations that may arise, the chosen methods will be effective.

Of modern information technology on electronic resources using a variety of software for the creation of professional education in the form of wide options. Thus, power point, macromedia dreamweaver, microsoft frontpage, html editor and microsoft word, adobe photoshop, CorelDraw, acrobat reader, Director of software such as macromedia flash options is great. For example, this program is macromedia flash 8 is an example of animasion. Animated izomeriya of complex compounds from using this program are prepared.

Of The Republic Of Uzbekistan's First President I Will.A.Karimov his “dream of harmoniously developed generation” in the game I have the following thoughts- “the age of modern science-technique, without access to the achievements of science, in general, stand, be the owners of them highly qualified conditions of the national economy of specialization such defect, we stand, we could not radical change” [7, - b. 525].

Biology science axisitishda testamentboro texnol thegof iyadan foydto alan, training effect,dorgaccounted for increaseda muhim factor h,christb,laned, I. "Biology" science-related topics in the classes of complex chemical compounds, electronic structure, izomeriya traditions in the study of phenomena'aviy when you go to get training in methods, the themeof the worlddevelopment difficult by the students to the introduction of pedagogical experience in the testing process are known. Therefore, this topicthe worldto read related to the implementation of information technology multimedia app has developed. Likely to invade computersьyüter sessions in the creation of most of the earliero, the process is likely to invade computers and electronic structure of complex compounds chemicalyuterdan mulьtiplikasiya was.

Kompьyusweat hobbies elektron hujjatiga 20 ta moval va quality issuesda Ibora test kiritildI. Mashg'ulot of the text did not go through the voice of the teacher explained. Monitor the processes in mulьtiplikasiyasini, that is, particular importance was given to explain the dynamic of the model.

"A combination of complex and chemical properties taken the theme of" likely to invade computersьyüter hobbies Universities (Qo'qondpi, NDPI, tdp)of 302 – teaching of chemistry of study of bachelor's students of the groupsida was conducted. Mashg'ulot shown in the screen through the details of videoko'z was made. Likely to invade computersьyüter disket likely to invade without all of the students through independent computers hobbies recordedьyüter workouts that will create the conditions to familiarize yourself with re. Kompleks birikmalarg tuzbe the set vis a synthesis ofzI itchflour Nobelь mukofotiga sazovor necklga scientist - A. Verner tsonrisingda data to bher husband atdI. Shvotsar kimyagari A.Verner 1866 , the year ofda Frytsiyag M,yulze chinahsettingsda tucountryilgan. Kompleks birikmalarg onchisi theirininig "Pomegranateglekarzemk kimya sohasidagI sidegI qarash the" kitobIda (1905 of the year) eng muhim kompleks birikmalarg synthesis ofzlto get summarizegan. In 1913 , the year ofda kompleks birikmalar sohasidagl muhim work itchflour Nobelь mukofotiga sazovor necklgan [84, - b.504].

In 1893 , the year ofda Verner kompleks birikmalark gimyaviy tuzilish norzof ariya will createb, em kdddinasion birikmalar deb itsdI. Bhe norzariyag the basis ofda atom bir-bir bwith its ownamong bocountrylanganda basic valek kuchsda tashfor thebee, duringpantscha valek kuchs bwith its ownamong birikworks waxkindegan economyoya laydI.

So needless combination of complex compounds, the central atom of its molecule or ion ion or if you have a few ions or molecules, i.e. the ligand is qurshab. Until now it is not given to a complex combination of a clear definition. Academic Yu.N.Kukushkin complex compounds has given the following description: “when we say a complex combination of crystals may be the case, the solution may be a combination of the ligand with the central atom in the structure qurshalgan available, you should consider” [64, - b.455].

Biology lessons in graphic organeyzerlardan very easy to use (table 1).

A conceptual introduction to the subject for teaching in complex compounds izomeriya schedule.

This method requires the comparison of the three compounds in more complex and will be particularly useful when izomeriya. The structure of the table is as follows: taqqoslanadigan izomer on vertical, horizontal comparison on this fulfilled is placed on the features and properties.

"The class of complex compounds of them be taken, and the role of chemical properties are the subject of" the read of innovative technologies "Pinboard" methods (1-methods)introduction to accounted for.

On the positive side of this method main, that is, in students of chemistry textbooks and training manuals transported to the formation of skills and qualifications to work with is a big help, perform independent oshwas ira.

Training of the units that divide the track take the stage 4 group, and each group of 4 units of the track is a complex combination of the type (asidokomplekslar, ammiakatlar, aquais acomplex and poligalogenidlar) studied from textbooks and manualssh certain types of formula to them and all the relevant equations and to write the cards to come in the same measure of knowledge is presented. Device is divided into 4 parts, each section will be recorded to the upper side of the name of the class of complex compounds. Reduced by way of the removal of the building from each group to the class of appropriate knowledge, the question of the subject of the formula and equations on the whiteboard attached to the order of chis iq. For this purpose, a "magnetic applikasiyasi"can be used. The criteria to perform the task given to the group put the points in a short time, the procedure on the subject of cards attached systemicshI, the complex of the formula of the equation of the reaction between them andng, inspired on set. Students interesting in doing this processshand I activity increases dramatically. "Pin ningboard" word "attached to the whiteboard" means. "Pin ningboard" and the nurturing side that develops methods of the advantages of the task. With this method students master the knowledge and skills of students and to maintaining logical and systematic thinking skills and skills be developed.

When each function is passed to another class of complex compounds group is replaced. For example, the complex compounds of the formula are attached to a team, in the second case , called the detail is attached. The usefulness of this technology sinceshI, full of didactic materials be prepared to work with the members of the group so far a clear procedure, the systematic implementation of the tasks entrusted and, especially, the knowledge of students related to the subject of education is an independent full oshirilganda are.

2-methods "brainstorming" method would be to the current group track group 4 of the training process. Track each group “brainstorming” to conduct four different content from the one particular using the method in question is given:

1) Modern electronic structure theory complex compounds how many comments you'll give?
2) Features of the electrolytes on the solution of complex compounds found nosebleeds dissosiyalanish consists of what?
3) Compounds in complex phenomenon izomeriya on how many will you give? Come by and write of their kind?

4) Come carry out unique functions of complex compounds in the human body is an example of how, in particular?

The answers given above are recorded in a separate poster of the questions from each group is going to write. As a result of participating in all the recorded responses page will be generated. Determine which of the answers is repeated. Depending on the completeness of the answer written on the right and, group 4 points the track is laid out.

"Complex compounds" in the subject to learn "brainstorming" that shows the results of applying the method of the effectiveness of the training is high.

Tavsifdagi innovative control tools. In modern conditions, not only in the organization of the educational process, but also in the control of the educational activity of students at the same time to have recourse to the description of the innovation used. None of the two in the same place of such a tool – educational portfolio and "Assesment" dwell on technology.

In recent years, teachers in modern education professional, while the improvement of the content of the student's educational activity, as well as on the basis of certain indicators of the quality of both the type of activities enough, unbiased assessment in relation to the need is increasing. Portfolio evaluation to the activities of teachers and students, to what extent the possibility of an indirect analysis that serves to provide quality and effective. Finally, the portfolio of them to the person of the teachers and the students with various activities examination (examination)gives you the opportunity to introduce the members of the commission are in the process indirectly.

Technology assesment of students knowledge, skills and qualifications to the level of comprehensive, unbiased reviews assignments package offering the possibility of it biografik survey, a description of accomplishments in the field of education, the students individual assignments, group discussions, interviews, creative work, tests, individual case study presentation, expert tracking, and business was founded as a role from the games.

The first chapter on conclusions

Human rights and freedoms of the republic of uzbekistan for compliance with the spiritual renewal of society, the form of the development of socially oriented market economy, a democratic state of law and civil society to join the community, providing open world of construction. The nation's rich intellectual heritage and human values on the basis of the modern culture, economy, science, engineering and technology personnel to prepare the achievements of uzbekistan on the basis of the formation of an important condition for the development of the perfect system.

1. In pedagogical higher education institutions “biology” read to achieve than a technological approach to science, teaching, information technology, electronic modules and electronic textbooks animasion applied on the basis of programs of the preparation of the topical issues considered.

2. “Biology” utilize technical knowledge of integrated science teaching pedagogical and information-methodical developed to supply. “Biology” science related to the recommendations of the lessons on the module, case study assignments, laboratory sessions and assignments to work out for both, will determine the necessity for the creation of didactic and methodological manuals.

3. Opportunities of information technology in higher education “Biology” science in the process of teaching the use of electronic educational technique that allows you to use the supply and production of life, “the teaching of chemistry” improvement of the science program of study bachelor of science add the latest achievements of the program focus on priority directions of development skills the main purpose of the curriculum, from science to come out of the block content is also exactly the same hours, and additional changes should be.

4. “Biology,” science technology and innovative teaching methods that serve from the use of information and communication technologies for development and introduction to the practice, “biology”, which serve to improve the process of teaching of science consider that it is necessary to pay attention to the following issues:

synthesis and achieved positive results on the basis of the development of the "biology" of science teaching in the information-educational environment of creation;

In pedagogical higher education institutions "biology" create a learning environment which allows the implementation of a methodological supply of innovative science teaching.

II CHAPTER. IN HIGHER EDUCATION INSTITUTIONS "BIOLOGY" SCIENCE TEACHING

§2.1. "Biology," science, principles of materials selection in

The science of biology and the study of them on the content of the principle of the choice of educational, didactic, as is known from the analysis of the scientific literature, the use of these materials in the educational process education from the aspect of improving the content and quality of educational work, students in theoretical knowledge, practical skills and skills to the formation of a wide creates opportunity.

The science of biology to create the content of the materials on the following problems must be solved:

1. The science of biology teaching and increasing the effectiveness of the selected materials in creating the content of lectures, laboratory and practical sessions in the traditional, determine the use of any form of unconventional methods.

2. Fikrlashlarini independent of the science of biology students on the content of selected materials, the development of abilities to know that constitute logical mushohada doing take into account.

3.The science of biology students on the content of the materials selected, the technique of teaching to develop a life of.

The science of biology related to the content of the educational process is understood about the use of the materials selected for the following purposes:

1.The selected material "biology" of the science curriculum and to be compatible with the requirements of the program, students mastered without help to get tormented.

2.Students constitute the form of the scientific outlook.

3.Independent and creative thinking in students grow up.

4.Abstrakt the thought of students, the development of logical mushohada doing partnership.

5.Possessing theoretical knowledge used in the practice of the students know.

6.The science of biology related to the content of the educational process on a systematic basis and be targeted in the use of selected materials, the identification and selection of necessary pedagogical situation.

The science of biology related to the content of the selected material, you can put the didactic following requirements:

1.The content of the materials selected must meet the requirements didactic and methodological;

2.The content of the materials selected simple, lo'nda completed and should be expressed in a short text;

3.The content of the material should be selected and the true facts;

4.Selected materials and test questions should be clear and understandable.

Creating the content of the science of biology and the results of scientific research conducted on the theoretical and pedagogical methods of teaching experience – practical didactic scientific conclusions and practical recommendations of the work testing the results obtained shall be subject to the following requirements:

1.Selected kontsepsiyasida of materials science biology, state educational standard, the curriculum I have found the program and its content.

2.Science – technology – production to the development of the modern answer.

3.The selected material does not repeat.

4.Students unfamiliar, difficult terms are not.

Based on the idea above, students in the science of biology related to the content of the selected materials, lectures, laboratory sessions and workshops according to familiarize.

Below are the materials selected, the pedagogical experience – in the testing process to test this methodology, we recommend that:

1.Biology the science of material selection on the content;

2.Put the problems according to the content of the selected material;

3.Put the take off of the problems, as well as the use of modern teaching methods;

4. The ultimate result of the problems found their own solution, the established goal has been achieved, the task has been completed.

The content of the selected materials on the subject of biology 2-are listed in the table.

§2.2. "Biology" on the subject of theoretical, laboratory sessions, demonstration and virtual forms in the process of performing an experiment to conduct integrasiyalashgan animasion program without the organization and methods.

Biology and developing new educational science in these areas because it is likely to invade the computers as the form ofeks to all known the same time the number of combinations noa lot of from a number of organic substances detected. Likely to invade computers atethe importance of the combination of ks bebe checking, they realized that participates in the process of plant and human life, hydro myetallurgik rare clean a mewho are likely to invade computerseis performed using a combination of ks. Kthis islikely to invade computers at pginaeks compounds of the medicinal substance. Most importantly, it new discoveries in this fieldchsurveys is the source of flour [112, 197-200-b.].

Biology practical experience term basis an experimental science is for him on theoretical knowledge appropriate to demonstrate the experiments to conduct, laboratory work performance and issues withdrawals on searching mastering can be. The same reason for Biology the scienceto teaching students with theoretical knowledge to give with along this knowledge to strengthen, deepen, expand , and practice with tie to laboratory work, the show is a experience, issues to take off and organize to and conduct of the educational process, the important components are. Biology the scienceto the teaching of the purpose, objectives and content, forms and methods are also improving has been. This , in particular, is currently a chemistry teaching process in the form, methods for a while modifications to add necessary that is known to be getting.

Conducted research in the process of such a conclusion we have come, you Biology sciencefrom lab work, a show experiences, issues to take off natural position in the organization to and to conduct the way if they choose, teachers with modern methods, the use of not knowing are left, you this lesson, the teaching of the modern virtual in the form of the organization to and to conduct the way if we choose, the teachers of the actual equipment-equipment with you to work, that is, the natural method is grown, the laboratory of chemistry teacher with his own two hands you're done, be held to follow, the results take the conclusion must be.

In our opinion, Biology the scienceto teaching in the context of laboratory work, a show experience, independent education students each, both in the method, that is, the natural way, virtual in the form of an integratedasiyalash the case in parallel to read up make will be implied effective achieve one of the ways is.

Theoretical and practical material, which consists of "Biology" science of knowledge when conducting a lesson on traditional methods for teaching difficult students by the introduction of the development of the pedagogik experience-is known from the results of the test. Because the computers are likely to invadeemay be formed of a combination of ks, structure, properties, izo Iek was rlarelish and function in the body it is hard for students to imagine the process will determine that. Therefore, this information is subject to explain the subject of tem applied to teaching xnologiyalarinietodikasi has developed. Developments in the lesson's text in the main part of the monitor berilgan and with the sound of the teacher are outlined. Especially, the subject of dynamic processes in complex modeli mulьtimedia embodied in the monitor and explain great attention to berilgan [114, 95-99-b.].

The monitor is likely to invade the computers inthe eks compounds on the doctrine of the founder.Verner porteoldest and largest embodied, scientists brief information about berilgan. A.Verner likely to invade computers createdeks a combination of theory on the structure of this comment koordinasionshattention will focus on.

In 1893 , in the year of shvaeyt a purple chemist, likely to invade the computers ateks compounds based on the teachings ofchisi, Dieelь award winner alьfred verner very much likely to invade computers ateks with a combination ofez made, their properties and the structure of theshaccounted for ekspoeriment by melearned with toddler [97, 487]. Taken ekspoeriment, based on the evidence likely to invade the computersand ein combination ks metall spindle atomicent let'schs yo'naldi all of its surrounding spaceshof the edge of the worldcha ted spread kisegan will put forward the idea. As a result, meg tall around the other atom atom atom orheruhI valent let'schs no more thanchin the amount ofshadi. This phenomenon is A.Verned koordinasiya rebkatom to the central atom, which show phenomenon oordinasiya deb, the spindle of the central atomeexcess from the number of the gardens and its additional ntent deb is called.

For example, $PtCl_4$·RAISES$_3$ or $[PtCl_4 \cdot 2(NHK_3)_2]$ neytral likely to invade computers atein the image the structure of the main spindle ksent line of interaction with additional valent interaction pointsIr-line shows. The main spindle platinument with ptcl$_4$ to form, in addition to being additional valent account on two more ammonia moleegets attached to the members of laugh. A.Vernerning likely to invade computers ateval result of compounds is formed in appointing ksepower on the idea of niche ntent connection elektron and quantum-mexa ugly theories (Lewis, 1916, the year of 1927. L. Threshold've N. Sidjvik) was created after it becomes apparent [85].

Soek, A.Verner $k_2[PtCl_6]$ likely to invade computers ateks structure is a combination ofshaccounted for the blind following of thecityat describes. Thus, the main spindle of platinument let'schI 4 units duringshimchspindleelet's ntch2 includes the members of chloride ions attached to I.

A.Vernerning likely to invade computers ateks compounds is formed ofshduring the central atom and the basic idashimchspindleent let'schs n anion oreytral costeplease laugh fastening members of the highsh process likely to invade computersьwiner monitoridan to the screen was down, it is a show. Shitek, A.Verner koordinasion embodied theory of the theory of the main text on the screen, the explanation of the process is done with the voice of the teacher. A.Verner according to the theory, any computers likely to invadeeks a combination of costem laugh in the central place mostly in a positive chargeewhich are likely to invade computers consisting of tall at ioneform ks a deb called occupies the central ion. Her around the ligand , as referred to, as well aschI the anion or neutral molecule places.

The central ion with the ligand by adding, the complex combination of Ichki of sfera form makes it. Ichki sfera square bracket Ichida are written. The central ion from away is located the ion complex of the outer sfera, the yield will be. A.Verner coordinatesasion the theory on the basis of $K_2[PtCl_6]$ (IV) likely to invade computers ateks a combination of the structure of thepicture you can show the following:

Potassium geksaxloraplatinat (IV) - $K_2[PtCl_6]$ the complex combination of the shhim o'xshash too many complex compounds chemical properties of tekshircityin the world in them ligand of a different texture to able that is, that a same 'sch , with the central ion to pull and typeshaccounted for shows. Therefore, main , and duringshimcha valence between almost difference will be, that conclusion has come.

Complex compounds formed with be and the structure of the processng mulьtimediasi the screen on the show to appear.

Given anima ofyathe future to retsIya we will explain on the basis of the equation. Iron (ii)-sulfate in the solution of $k_3[Fe(CN)_6]$ a less amount of red blood salt from a solution, I have added a complex combination is formed:

$$\text{At } fes_4 + K_3[Fe(CN)_6] = Fe_3[Fe(CN)_6] + K_2SO_4$$

But this theory using the basis of the law on the central Kulon is found between the experiment detected by the ion with the ligand gardens garden garden energy is the energy difference. This theory of a complex ion is formedshthe nature of the central atom of the nest, electronic composition bog'liq explain the magnetic texture of the complex would beydi.

A.Verner's theory and in determining the nature of the basic and additional koordinasion L.Threshold've N.Sidjvik name is important. N. Among donors with the central atom in ligand complex compounds sidjvik-aktseptor comes to the idea that interaction creates a connection. Complex compounds chemical connection in quantum-mechanical too complex to interpret donor-valent and links aktseptor be formed by the mechanism on the basis of methods, structure and texture of their images.

Copper(ii)-tetraamingidroksid $[Cu(NHK_3)_4](OH)_2$ formed a complex combination of st structure review and theqwe are at.

We will explain the equation given on the basis of animated reaction. Copper (ii)-ammonium sulfate solution, alkali solution, I have added kontsentrlangan less amount of copper (II)- dilute alkali is formed.

$$Cure_4\text{+}RAISES_4OH\text{=}cu(oh)_2\downarrow\text{+}(NHK_4)_2SO_4$$

Less is more, I have added ammonium alkali solution, the precipitate obtained transparent purple-colored copper (ii)- complex compounds of alkali tetraamin a solution is formed.

$$Cu(oh)_2\text{+}4NH_4OH\text{=}[Cu(NHK_3)_4](OH)_2\text{+}4H_2O$$

Copper(ii)-tetraamingidroksid as nosebleeds is to decompose on strong ion electrolytes.

$$[Cu(NHK_3)_4](OH)_2\text{=}[Cu(NHK_3)_4]^{2+}\text{+}2(OH)^-$$

Copper(ii)-tetraamingidroksid $[Cu(NHK_3)_4]^{2+}$ complex ion and the copper atom in the complex to be formed to consider the process of Cu^{2+} ion in the formula, we will write electronic.

1 $cure^2\ 2s^2\ 2p^6\ 3^2\ 3p^6\ 3d^{10}\ 4s^1$- 2^- - $Cu^{2+}\ 1^2\ 2s^2\ 2p^6\ 3^2\ 3p^6\ 3d^9\ 4s^0 4p^0_x\ 4p^0_y\ 4p^0_z$

The expression of copper ion valence orbital appreciate it.

$3d^9\ 4s^0 4p^0_x\ 4p^0_y\ 4p^0_z$

↑↓	↑↓	↑↓	↑↓	↑				

Copper (ii) tetraamin gIdronks complex texture paramagnit bow downa. It 3d orbof Italianmarketing andI tothe e - vujidgbring forth adI. Therefore, gibridlanishda 4svparticipates in a 3p orbital.

S and p orbital 3 in 1 copper ion, which is formed from them, as well as 4 units sp^3 onbsince it was empty orbital ridlangan $[Cu(NHK_3)_4]^{2+}$ complex ion b the yield ofthislish in donor-aktseptor aktseptor performs the function of copper ion engine.

In the ammonia molecule, which is the ligand of the complex H_3N: taqsimlanmagan e-pair donor has the function of the electron. As a result, the cu^{+2} ions in the ammonia molecule, the donor-engine aktseptor mutual impressiveb copper (ii) tetramin $[Cu(NHK_3)_4]^{+2}$ complex ion forms.

In the second example xrom (III) geksaakvaxlorid $[Cr(H_2O)_6]cl$ are my_3 complex combination of hk osil would bethissits stqwe are at. A combination of complex crystals oktandr has the form. Complex compounds are formedshida Cr^{+3} ion in d^2sp^3 gibridlanishsh happens.

Cr^{+3} ion with a water molecule ta'sirlashib, $[Cr(H_2O)_6]^{+3}$ ions creates. The water molecule, the oxygen atom has electronic taqsimlanmagan pair. Pair with the oxygen atom of the water molecule in the structure of electronic taqsimlanmagan 6 bosh gibridlanish d^2sp^3 orbital among donors-aktseptor thiscomes into being as a result of the effects of koordinasion tied to zaro. As a result oktaedr ge in the form ofakvaxrom (III) $[Cr(H_2O)_6]^{+3}$ complex ion is formed.

In the center oktaedr Cr^{+3} ion, water molecules, while nodeslplaces many years ago. The complex ion is formed in the process of gibridlanish both the central ion, the ligand orbital to come gibridlanish to bind the actions of the crystalline grid of the structure of the complex is animated. The nature of chemical bonding in complex compounds, that is, A.Koordinasion about the structure of complex compounds of verner, elektrostatik to draw Kosselning forces Sidjvikning donor-valent connecting aktseptor to be formed on the threshold of the complex mechanism explanation of the theory and methods of hand-written text as it is written in giving voice to text is written and explained.

In addition, the combination of the number of formed in a complex combination koordinasion and its structure is important, you also learn about the complex combination of knowledge about the ability reaksion you will be able to. The concept of koordinasinon give information on the number of complex compounds is important. Depending on the number of koordinasion you can freely give information about the complex structure of (3-table).

Research into the reactions of the metal complex in the work, etc. have been studied. The structure of the obtained complex compounds can show to students through computer freely and analysis skills to synthesize their thinking and shape.

The complex compounds izomeriya T-graph comparison using.

The t-graph of the debate at the time the twofold answers (yes/no supporter/against) or comparison-answers to write organayzer universal graphics is conflicting.

T-izomeriya phenomenon in a complex combination of using the graph can be compared.

Izomeriya structure of options in complex compounds	**in a complex combination optik izomeriya of options**
depending on the type of Izomeriya structure of a complex combination of the change in	position is to track changes in the spatial

Complex compounds are well known to be achieved in each specific suyuqlanish boil and the temperature is distinct from a simple substance. All of which attracted the attention of researchers in the color, electrical conductivity, oxidation-reduction properties of colorful, magnetic and other properties includes [62, 432]. By checking the change of color is often a complex form in the system is stable or unstable and his combination of ingredients can determine that. The complex study in infrared light through a combination of content that would win the inter-atomic contacts of the description you can get to know. Kvanti be in the form of the complex that have been the length of the waves will cause you to change the color of(4-table).

As a general concept used in this report, giving information from a table in the text. In the process of conducting laboratory classes obtained from the reaction then compare the color of the complex, analyze is achieved.

Theoretical knowledge, laboratory work, and demonstrated experience in the form of natural and independent virtual organization and conduct of the learning is done in the following two different methods:

first, in natural form - theoretical lessons in the classrooms is carried out;

second, in the form of virtual – computer software mobile app or through the network in the process of performing and creative independent reading assignments was carried out.

Higher education institutions in "biology" in the process of teaching of science, theoretical exercises, laboratory work, and demonstrated experience in the following 5 natural and independent virtual form of teaching method has been developed without integrasiyalashgan:

theoretical study of the materials of the first lesson.

the second study of occupation using computer software.

third - occupation in the form of virtual chemical laboratory of chemical agents in some cases when failure is a necessity to calculate and the results of the performance manufacturing experience, the production of relevant evidence;

the fourth lesson of control utilized in knowledge, self-control;

fifth - training process to complete an education that is relevant to discuss the implementation of education students get on konsultasiyalar and independent from the teacher.

As an example of application of the above methods, exercises and laboratory work in its virtual form read more to consider.

N laboratory worklaboratory work provides the theoretical part of the process of the azar study materials and the performance of the work necessary to complete the laboratory equipment, the purpose of the work, that the work is complete and the stage of the procedure, after a certain charge to do the work to advance their knowledge of scientific-theoretical issues relevant to the content of the information revealed. Information, text, graphics and multimedia can be given in the form of. As an example below, "and the combination of the texture of the complex is taken" is given in the theoretical part of the laboratory work (2.2.1-picture).

Subject: the science of biology, its functions. A.Verner's theory of the structure of complex compounds.

Subject aim:

Educational: the science of biology, its functions. A.Verner's theory of the structure of a complex combination of students to generate new knowledge on.

Educational: the science of biology, its functions. A.The theory of the structure of a complex combination of verner to generate new knowledge on the historical data deliver.

Developing: students in the science of biology, and its functions.Verner on the theory of the structure of a complex combination of sources through knowledge increase.

Pedagogical objectives:

The science of biology, its functions. A.The theory of the structure of a complex combination of verner give general information about;

The science of biology, its functions. A.The theory of the structure of a complex combination of verner to generate new knowledge on.

The results of the educational activity:

The science of biology, its functions. A.The theory of the structure of complex compounds of verner will have general information about.

The science of biology, its functions. A.Verner's theory of the structure of a complex combination of the updates you will know.

Type of training: education report, software applications animasion, group discussions, Blits –request.

Teaching forms: Public, collective and individual

Pedagogical technologies used:

- developing technology training;
- information technology;

Teaching tools: likely to invade computersьyüter, screen, proektor, such

The structure and distribution of sessions time:

1. The organizational part. 3 minutes

2. Lesson plan and goals. 5 minutes
3. Synthesis of knowledge of the content of the lesson and students. 68 minutes
4. Strengthen. 2 minutes
5. Home task. 2 minutes

Lesson plan

I. the organizational part**:**

a) define management access;

b) to prepare the lesson, define device, chalktechnical means.

II.Go, of course, be

1. The science of biology, its functions.

2. A.The structure of complex compounds vernersh theory.

Base phrases: complex compounds, its functions, A.Verner's theory, the structure of complex compoundsshI

PLAN:

1. The science of biology, its functions.

2. A.Verner's the structure of complex compoundssh theory.

The monitor is likely to invade the computers inthe eks compounds on the doctrine of the founder.Verner porteoldest and largest embodied, scientists brief information about berilgan.

A.Verner likely to invade computers createdeks a combination of theory on the structure of this comment koordinasionshattention will focus on.

Alьfred verner very much likely to invade computers ateks with a combination ofez made, their properties and the structure of theshaccounted for ekspoeriment by melearned with toddler. Taken ekspoeriment, based on the evidence likely to invade the computersand ein combination ks metall spindle atomicent let'schs yo'naldi all of its surrounding spaceshof the edge of the worldcha ted spread kisegan will put forward the idea.

III. Synthesis: the science of biology, its functions.

A.Verner's the structure of complex compoundssh theory on the subject of the knowledge obtained will conclude.

IV. Lesson reinforcement:

Control questions:

1. How are complex compounds that combination is said to?
2. What does it mean for the number of koordinasion?
3. Chemical connections between the central atom of the ligand explained how?

V. Home task: the science of biology, its functions.

A.Verner's the structure of complex compoundssh theory.

The commission for independent education:

The science of biology, its functions. A.Verner's the structure of complex compoundssh theory.

Recommended has been literature:

1. N.A.Parp, X.R.Head., A.G.Muftaxov inorganic chemistry. - Tashkent: Uzbekistan, 2003 .

2. B.D.Berezin, T.N.Lomova. Reaktsii dissosiasii kompleksnix soedineniy. - M.: "Nauka", 2007. -278 s.

3. V.V.Skopenko, A.Yu.Tsivadze, L.I.Savranskiy A.D.Garnovskiy. Koordinasionnaya ximiya. - M.: IX "Akademkniga", 2007. -487 s.

Topic: "be taken, and a complex combination of properties" development of laboratory

Currently, higher education, Biology science for mastering the theoretical and practical knowledge of the use of information technology produces good results.

Students are asked to give many comments from experience, thoughts and experience will be a continuation of summarizing is shown. The screen is shown in the reaction in equation (2.2.2-picture).

The purpose of the laboratory work in the form of videodars the equipment needed to complete the work order is introduced. In particular, the aim of this laboratory work is a complex combination of that and learn that, the necessary equipment vismut a solution of nitrate of potassium solution yodid, shtativ, pipetka, probirka consists of the work in order to complete the necessity to follow the rules of safety, probirka on the set, from manufacturing of chemical agents you want to use onigi noted, and the work is displayed in a way fulfilling automated. Complete laboratory work in this case, you should pay attention to the color of the solution it is said to change depending on the color reaction after the date of the complete forms of the reaction process is the analysis, the conclusion is said. The reaction equation of the computer screen will appear. In this laboratory will help students develop the knowledge and skills necessary to complete the work independently.

Part of the control of the laboratory work laboratory work and theoretical materials of the laboratory experience consists of knowledge of the complete process utilized in a control test. Utilized knowledge of the virtual laboratory for the purpose of consolidation and check the "current control" and there is access to the part related to the work of the laboratory by performing relevant tests, teachers can check their own knowledge and consolidation (2.2.3-picture).

2.2.3-picture. Complete the test page.

Also, the control part, the students laboratory work, independent work, along with demonstrations on experience with educational materials on the study of complete self-control will go. Students doing independent work of higher education institutions in the educational materials given in the form of a virtual portal to learn from the theoretical aspects and self-control, to the advance of knowledge, creative exercises, assignments performs. Self-control in the current control shall perform creative assignments relating to students. Control the current control module on successful past students from doing the work have the whole space.

During sessions in conjunction with innovative technologies and does not seem to show the students get the desired information from the computer of sessions, the efficiency is further increased.

First, the students in computer classes similar to the experience for termination of chemical processes chemical knowledge to bring more high adopters tafakkurlash imagine seeing with their own eyes; and secondly, the structure of complex compounds in the process of computer training of students multiplikasiyasiga interest is high.

Indeed, the only one not to be taken on the topic of a complex combination of chemistry, but other topics in the teaching of science in the modern information technologies in the minds of the students no doubt be the best assistant tool to break. Read literature to understand many of the internal structure of a substance, you will need to learn in the case of thought. The theme of the level by using the form za teachers students a thorough understanding of literature from the consolidation will facilitate more if I remember to read independently. More convenient to take lessons using the form za is an indicator of higher quality and most importantly effective.

Many electronic devices used in the process of education is currently very prepared. Modern electronic devices are programmed. Training in the use of information technology facilitates the study of the science of biology[23,45; 82, 24-27; 96, 256; 107,1000; 133].

This increases the level of mastering the subject of student learning. In addition, this process of mastering depending on the level of students to teachers, students to teach: techniques helps in the formation of extra questions on the subject.

The task of the situation.

1. A case statement. Students do experiments on a complex combination of the bumps. To get coca complex compounds$_2$*$2H_2O$ xinazolin of substance of substance(ligand), and the mutual experience complete assignments as given. But Nadira ligand in water, while they do experience Gulnoza spirtda melted. As a result, the sediment balance in the same experiment nodiraning like crystals in the experience of gulnoza in case the substance is formed.

Case study question: comment of chemical processes?

Case study analysis:guidelines for students:

1. Read carefully the information given in the source.
2. We should seek understanding of the difference.
3. On the basis of the data utilized to create your option to answer a question.
4. Answer options response and overall make your small group discussion with other representatives of the form of the option.

Take yours off process. Read carefully the information given in the source, and understanding the difference provides the option to answer a question on the basis of data utilized to create your small group with other representatives of the answer options were discussed and the shape of the overall response option

Case study solution: (options students).

The solution of the teacher: 1. The properties of the complex and the ligand in the process of being formed in each of the feature you need to give attention to achieve. Well spirtda xinazolin than water melts, suggest that this complex may be formed in a combination. Complex compounds, the central atom ($Co^{2+)}$ and standing around him qurshab ligand (xinazolin, chloride and water molecules)is a complex substance composed of s.

On laboratory assignments.

Students proficient laboratory sessions reinforce the theoretical knowledge, skills and qualifications of the shape. In laboratory sessions, students should there be a business notebook. The purpose of the training, laboratory equipment, educational equipment, laboratory teach students: student assignments be worthwhile to come the sequence of activities and information about hobbies. In addition, in laboratory sessions, students must follow the safety rules.

"Be out of a complex combination of" on the topic of assignments

1. Case description: in the rounds xlorofil plays a big role in the process of photosynthesis. It is important in the process of development of plants and living.

Case study question: I want to go complex compounds Which the word about how you do it if you comment?

2-set the task. "The balance of a complex combination konstant on the subject of your assignments

A) what constant means the balance of a complex combination of?

____

V) how are different? ______________________________________

S) complex compounds is any different from other substances?

____

3-case study task

You'll give any comments to the following process? Explain how the body of knowledge on the subject of analytical chemistry mainly your color, will you?

Fe^{3+} of $[FeF_6]^{3-}$ and $[FeCN_6]^{3-}$ complex ion energy diagram [91]

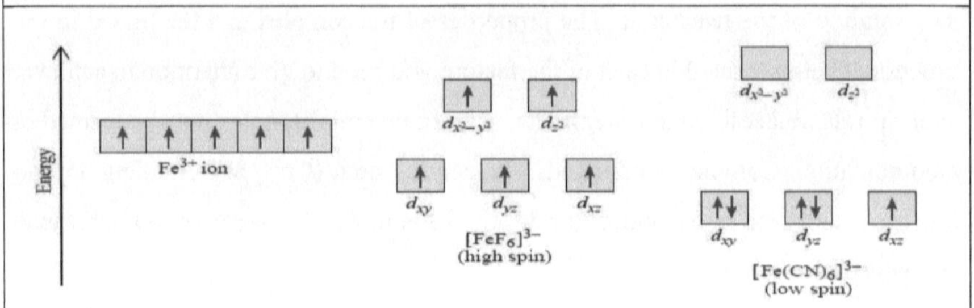

Tmetal ions in uzlari. *From leftto right, looking: the oldest and largest^{3+}, Cr^{3+}, Mn^{2+}, Fe^{3+}, $Co^{2+}To^{2+}$, Cu^{2+} ion, the color of the solution.*

Amid the donor atoms in the structure which keeps acid carbon, nitrogen, metal ions in the structure with a combination of some organic compounds which are prone to form and keep koordinasion [121]. In addition, amid amid properties and electronic structure of a mixed group and directions for many reasons, very interesting. First of all, this group of elements as the main components of natural, synthetic, enters into the composition of biologically active compounds. As it is known, the nature of the metal complex and depends on the nature of the group itself karboksil in different circumstances and to a certain extent and can koordinasiyalanishi koordinasiyalanishida amid the water molecule to the metal ion may be to compete with.

Problems such as donor atom of the ligand metal complex form for a student to take off as a combination of features you can choose to have. There is the importance of metal that have been in the development of the plant. In case of combination they koordinasion in plants is available. The functional groups in the structure of the active compounds koordinasion to be able to be more active again, not the level of the poison will be less.

Quantum chemical calculation method using marganes in scientific work at the present time (II), cobalt (II), nikel (II), copper (ii) and zinc salt of the electronic structure of the complex formed with the skills and xinazolin reaksion the analysis of heat to be generated, the electronic energy network detection [20]. The donor atom is the central atom depends on the nature of the siege without its uniqueness is shown. Xinazolon-4 tautomer in solution are observed to be in the form of 3 types.

1***H*** (**a**), 3***H*** (**b**) OH (**c**) tautomer can be see:

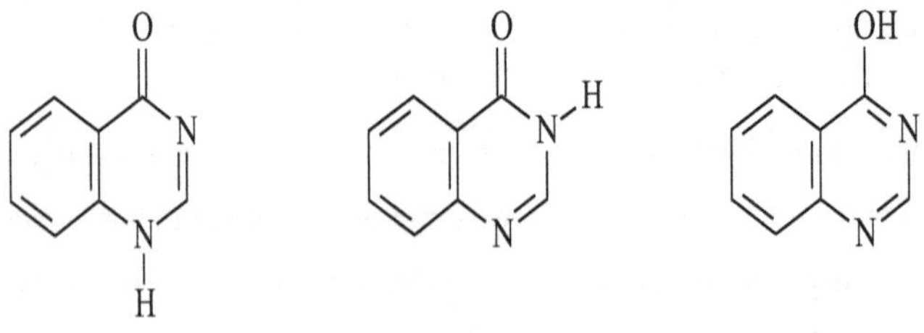

a b c

Xinazolon-4 and copper (ii) chloride is the salt obtained from a combination of crystals koordinasion do their research has shown that koordinasion combination of Cu (II) ion to two 1***H*** tautomer of the ligand in the form of (**a**), two water molecules and two chloride ions is formed with the oktaedral koordinasiyalanishi. Koordinasion out in a well-formed combination, center fold out well, fold in the center of cu (ii) ions in the lies. Taken koordinasion the combination of x-ray method on the basis of the results obtained in the molecular structure difraktsion 2.2.4-are listed in the picture.

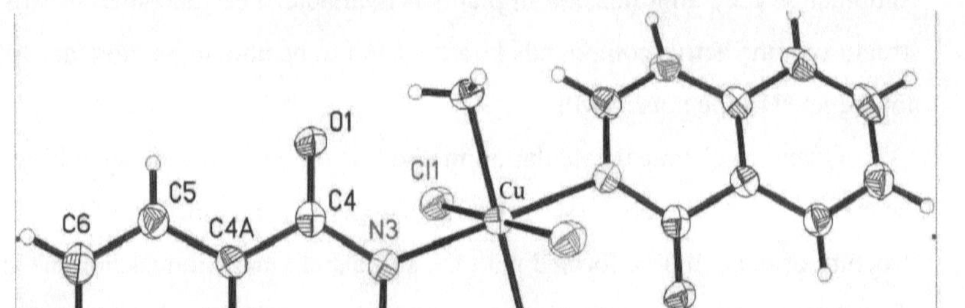

2.2.4-picture. [Cu (I)$_2$(H$_2$O)$_2$my cl$_2$] koordinasion the appearance of a combination of molecules.

Xinazolon dm-4 ligand, the cu—N_3 koordinasion relatively minor deviation of ligand into the garden with the plane and the Cu—N_3 the garden of vector (000) make up the corner.

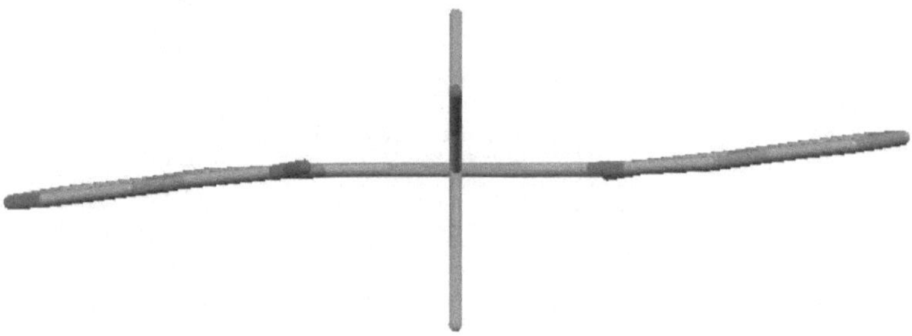

2.2.5-picture. [Cu (l)$_2$(H_2O)$_2$my cl$_2$] a complex combination's in plain view.

Internal structure of crystals and molecular analysis shows that water is also taken koordinasion combination koordinasiyalangan molekulalararo is involved in hydrogen bonding. Bound internal water molecules and molecular hydrogen xinazolon-ligand is observed between the group of 4 karbonil. Gardens with chloride ions of the water molecules that form hydrogen molekulalararo happened. Crystals of the cells *c* of complex molecules in addition to water and chloride ion ligand standing on the axis of the garden may generate hydrogen as a result of the *c* - axis chain bog'lanishli hydrogen formed.

In the structure of complex compounds xinazolon-4 ligand and chloride ions of the n—H...o and N—H...cl hydrogen to form a garden in my view link. Hydrogen links *a* - axis is observed, as a result, the form of chains (2.2.6-picture).

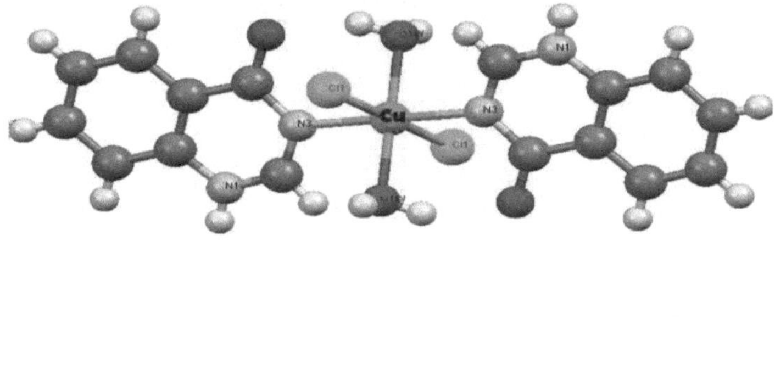

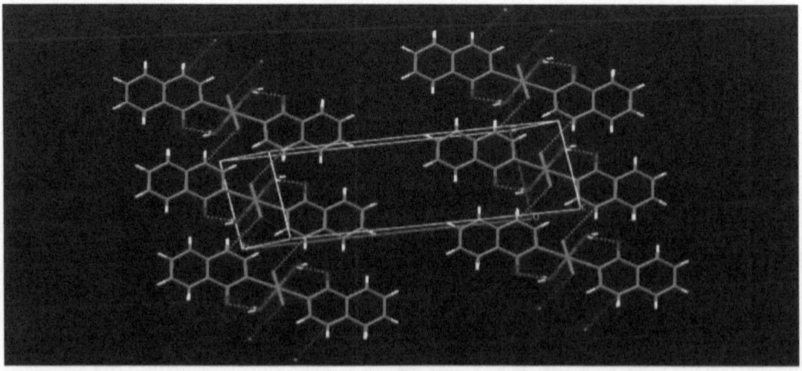

2.2.6-picture. [Cu $(L)_2(H_2O)_2$my cl_2] the structure of complex compounds.

The following 2.2.7-in the picture, copper (ii) based on xloridi room (a) and high (B) reaction of the product obtained at a temperature difraktogrammasi is shown.

Shows the results obtained, the conditions with the change of the crystalline structure on the basis of ligand and copper xloridi xinazolon aniklash the new structure, which is different from previously combination which can be formed since koordinasion [113, 1-9-b.]. Also, the reactions of the metal complex have been studied in the process of access.

Topic: complex compounds of izomeriya

Objective: to give information about the event and its types izomeriya of complex compounds occurring in the teaching of information technology support (2.2.8-10-picture).

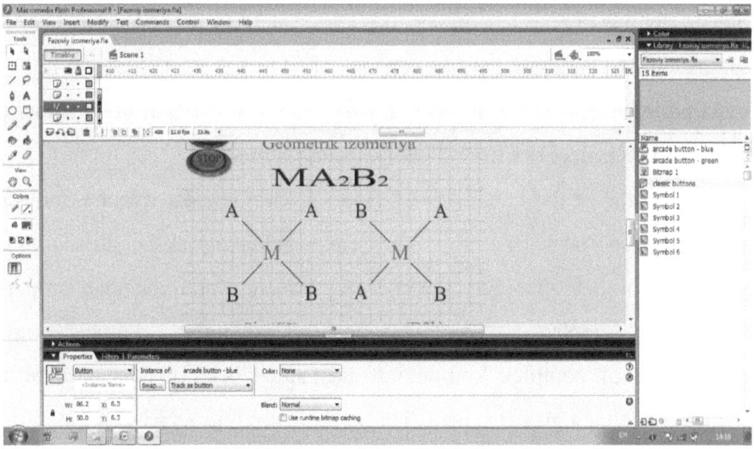

2.2.8-picture. Geometric prepared macromedia's flash working izomeriya the appearance of the window.

Such privilege or the phase of the image in flash macpomedia dacturida prepared animasiyalarinig the physique you need with reliable and accurate action, it will show 3d images. As well as complete the work during the period of the dissertation the structure of complex compounds, and spatial optical izomer so Macpomedia flash program was created.

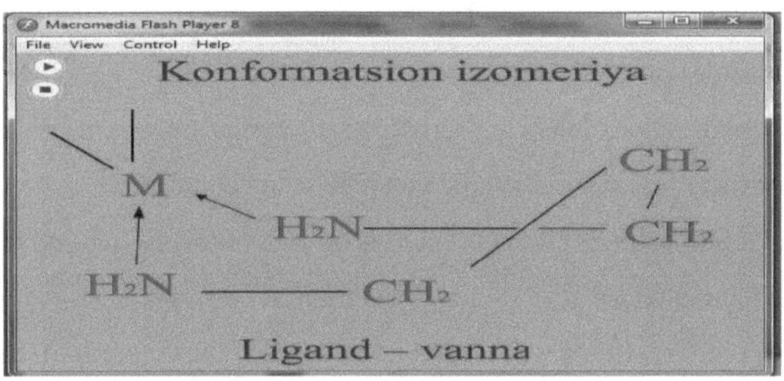

Ispring_Suite_6 doing the work.2.0 program was used. In the creation of electronic textbooks in the preparation and presentation of this program is very important for chemists. In this research work on the study of their properties and a complex combination of a number of works carried out, [113,1-9-b.; 114, 95-99-b.].

The complex combination of body position and understand the importance in science and technical the use of information technology.

The status in the chemistry of complex compounds, science, industry and describe the importance of living in their class asidokomplekslar, akvakomplekslar, ammiakatlar, poligalogenidlar, the complex structure and chemical properties of metal organic complex polikislotalar taken and explained. Complex composition and structure of complex compounds has since been understand the use of information technologies for them. This combination, in which the function of the body, the medicinal properties of modern science and technology, as well as may be used as stimulants in order to determine how the composition of complex compounds are studied.

Biology the science of information technologies on the basis of the arrowitish methodology of the topic to conduct research on theoretical and practical knowledge deepening, the science of biology thisrga factors can be familiar with modern techniques and technologiessh, practical and professional research kompetentsiyalarni formin order sh has conducted research.

The lessons pass in the process of active development of the students thought that knowledge can have an effective impact. The following question to the attention of students through the slides problematic tool.

1. Complex compounds of iron ion in the activity of the human organism on the basis of the importance metallorganik will show you how many comments?

2. On the importance of sodium and potassium ions in the nervous activity of the human body do you know what?

3. Iron drugs by unesco put forward the idea of what caused the creation of the store to come out?

4. In the structure of the cobalt (III) complex ion in the body, which will keep metallorganik what do you know?

5. Enhances the activity of complex compounds of action of how athletes?

The first question was solved by students problematic follows: the first two iron ion in the structure of the hemoglobin of students and that is the answer to the question is problematic. The second student from the fruits of the plant, apples, grapes, and nuts in them passes into the body through the consumption of which is the element of iron, that symbolizes the idea. The third student, which is a combination of the complex human hemoglobin in red blood cells iron is delivered to tissues that is oxygen in the body explains. Teachers on the basis of these solutions gives the complete solution of the question is problematic. Hemoglobin is the substance which is the cause of many to conduct her research. To look at the information given when the hemoglobin of more than 200 scientific articles will be published each year in the world.

A protein molecule is formed from a chain 4 units polipeptid nature globo. 1 globinning gems in each chain will be. Forfirinning gems and chain consisting of a combination of iron ions. Forfirin of iron ion in the center of the ring, the ring is in the color red because of hemoglobin forfirin. Have an empty orbital in which the iron element, if the oxygen molecules bind if oksigemoglobinga becomes. Oksigemoglobin oxygen supply to body tissues via the blood, giving dezoksigemoglobin to the lungs in case of returns. The structure of hemoglobin in the movement of^{2+} ion than the other combinations of iron is stable.

Structured questions for the topic also other similar problematic solution. For example, 5-metallorganik is dedicated to the importance of a complex combination misning problematic question. As it is known, the activity activates the action of the athletes efidrin called stimulants. Therefore, in sports competitions in the world efidringa analysis is conducted. It will form a colored complex with copper ion. Colored as a result of the reaction of copper (II) dilute ditiokarbamid is formed. This control is used the texture from dopa.

In the process of the lesson with this problematic question be solved on the one hand, a complex combination of assimilated knowledge about the importance of second hand, opens the way to the study of a class of complex compounds. Each incoming class of complex taken appropriate write on the whiteboard understand the chemical properties of the reaction equation, and is displayed to show interesting chemical experiments on them.

Therefore, the training process in the introduction of information technology to students in regard to the knowledge of interest related to the subject, an object, increased the effectiveness of a lesson, and students out of the development of knowledge by the introduction of high [119, 256].

Know the development of the educational activities of students on the module of complex compounds (case pack, laboratory assignments)

"The combination of the texture and the complex is taken" on the topic of laboratory

Lesson goal: students taken on the texture of the complex and the combination of knowledge, skills and qualifications formation.

Type of training: analysis-synthesis, educational games, work in groups

Training methods: practical, computer software.

Repository lessons: the manual on the basis of the experiments in multimedia applications.

Have to go training:

I. The organizational part. The student will be attracted to the attention of the subject is determined and access management.

II. To strengthen the theme.

1-group. To comment on the following concepts.

Concepts	**Concepts a note of**
Complexcompounds, s is	
the central atom of the ligand	
Koordinasion number koordinasion level	
Akvakompleks	
Asidokompleks	
Neutral complex	
izomer Ts	

Trans izomer	
complex Poligalogenidli	

2-group. Comment of the following processes:

T|p	the process	of the process of a note
1.	understanding about complex compounds	
2.	the naming of complex compounds	
3.	chemical properties	
4.	check out of the work area	
5.	the role	
6.	news	

III. Lesson description.

Students are given a complex combination of methods to get the theoretical knowledge about. Animated thematic created is displayed. In this animated gives me a complex combination of color change in a chemical event from the process.

IV. Strengthen.

"Be out of a complex combination of" on the topic of assignments

The topic carefully read the instructions and do as follows:

1-assignments. Case description: in rounds xlorofil plays an important role in the process of photosynthesis. It is important in the process of development of plants and living.

Case study question: I want to go complex compounds Which the word about it, how many comments you'll give?

2-assignments. Fill out the table and we should seek the definition of complex types.

T/r	for complex types	the specific features of complex
1.	The complex anion	
2.	Kation complex	
3.	Neutral complex	

V. summary and evaluation Lessons.

Topic: lectures on the subject of complex compounds of izomeriya lessons, the lesson development.

Lesson purpose: izomeriya of complex compounds on the subject of students ' knowledge, skills and qualifications formation.

Repository lessons: multimedia applications on the basis of the experiments in the manual.

Type of training: analysis-synthesis, educational games, to work in groups.

Training method: practical computer software tools.

Have to go training:

I .The organizational part. Identify and access management it is said to give attention to the lesson.

II. To strengthen the theme.

1. Comment the following concepts:

Concepts	Concepts a note of
Complexcompounds, s is	
the central atom of the ligand	
Koordinasion number koordinasion level	
Akvakompleks	

Asidokompleks	
Aminokompleks	
Roligalogenidli complex	
complex anion	
kation a complex	
Neutral complex	
izomer Ts	
Trans izomer	
the optical izomer	

IV. Sessions's go.

"Biology" textbook on the basis of experience in the electronic multimedia applications have been applied.

IV.Consolidation and valuation.

Blits-survey questions:

1. Izomeriya what is it?
2. It is observed that complex combinations of multiple izomeriya?
3. How complex that is said to akvakomplekslar?
4. The molecular structure of the complex ion is, how complex is it?
5. Asidokompleks what is it?

1-task case study: comparing a complex combination of processes that night in get:

Process of the	Central atom	in the ligand
izomer of Optical		
Geometric izomer of		

the spatial izomer		
Koordinasion the izomer		

Assesment answer questions during the control students in creative thinking at the same time by using technology to research, meditation achieves mulohazali.

Test	**The problematic situation**
the substance which carries out photosynthesis - at xlorofil A) a combination of koordinasion of cobalt; B) a combination of misning koordinasion; S) of a combination of kadmiy koordinasion; *D) koordinasion its a combination of magnesium deficiency.	That $[Co(NHK_3)_6]cl$ my_2 and $[co(NHK_3)_6]cl$ are my_3 give one of which is stable from a complex combination basing.
Chemical tracking(character)	**Practical skills**
izomer chemical is determined using a texture vary from one experiment how?	in izomeriya is based on a complex combination of the kind of event?. Of their kind come by, izomer of write.

Science biology module organization of independent educational methods.

The subject of biology test questions.

1. Izomeriya of complex compounds based on each formula and write the corresponding numbers under to determine the table.

1) koordinasion; 2) ionlanish; 3) gidrat izomeriya;4) koordinasiyali; 5) contact; 6) substitute; 7) the ligand.

FeCuF $6H_2O$	**Cu_4 $5H_2O$**	**H_2WO_4 $3WO_3$**	H [BF_4]	**Cure(SO)$_4$**	**Cure[Cr(C_6H_5)$_6$]**	**K[$BrCl_2$]**

javobli many non-standard test answer follows.

2	**3**	**6**	**1**	**7**	**5**	**4**

2. Write the corresponding numbers for a given substance.

1) 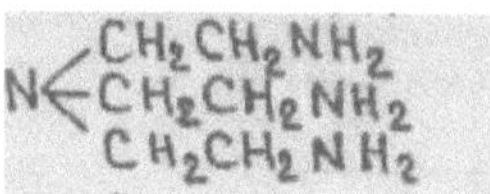	**complex parts**	**, the number of**
	Monodentantli	
2) 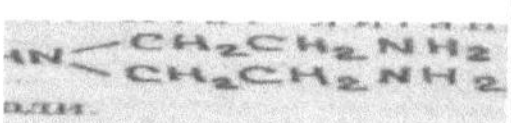	*Bidentantli*	**2**
	Tridentantli	
	Tetradentantli	**1**

3. Don't juftli complex compounds with their properties.

1	Akvakomplekslar (anion)	o f a	departure molecular structure of h$_2$N-CH$_2$-NHK$_2$ participated in groups such as complex combinations ofto say.
2	Asidokomplekslar (molecular)	B	at departure the molecular structure of C_5N_5 participated in the complex of ligand compoundsto say.
3	Xelatlar (cycle form ofch,I)	C	the molecular structure of the complex anion departure participated in compoundsto say.

4	the cyclical ligand	D	of the ligand water molecules

Answer :	1-	2 -	3 -	4 -

Answer:

Answer s:	1-D	2-C	3 - B	4 - A

4. Of the thoughts which are given below right? Schedule write the words yes or no answers.

A. Akvakomplekslar is in the structure of water molecules.

V. Complex xelatlarning represent the most common broadcast.

S. Free of complex compounds in the structure of molecules is available.

D. A complex combination achchiqtosh.

E. Complex compounds are used in pharmaceutical

J. A complex combination of salt.

1	2	3	4	5	6

1	2	3	4	5	6
yes	no	yes	yes	yes	no

5. Depending on the ligand structure of the complex compounds are granted to what?

A.*Monodentantli. B.C. k poliligandlithisligand p *D. Bidentantli.

6. Find the words leaving was going?

Thisligand quickly alma zshwho can rise complex–_________, hard almashhead raised while–___________are called complex.

A).* Lips, onl B). Onl, lips c). Ion aqua D). Xelat, molecular.

The topic carefully read the instructions and do as follows:

6. **Qo'sh salts and complex compounds venn diagram and fill out the table on the build.**

Qo'sh salts**to specific features of**	complex combinations of**features specific to**

8. Shown in the following picture and was not part of the complex.

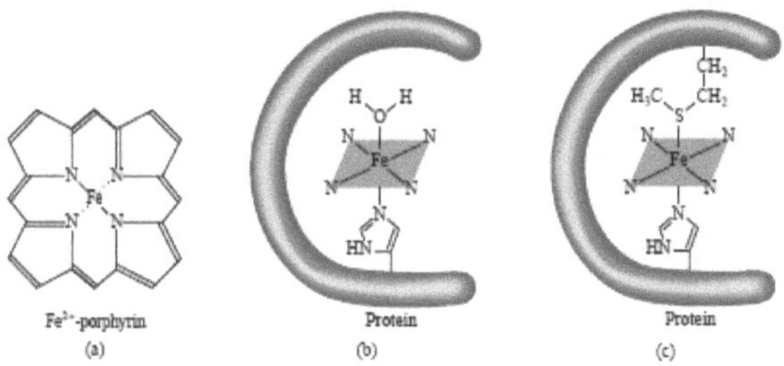

____

A - ______________________________

B - ______________________________

V - ______________________________

G - ______________________________

9. "The balance of a complex combination konstant on the subject of your assignments

A). What is the constant balance of complex compounds?

____

V).How different are? ______________________________

S). Complex compounds is any different from other substances?

____

Modern tools of educational materials are provided for educational literature, textbooks, textbooks, manuals, textbooks in electronic form, the electronic version of the text of the report, the electronic textbook with the essence of the concept as an integral link to get an accurate description of many of them worthwhile.

The electronic textbook – teaching methods based on the use of computer technology have designed a teaching tool if getting an education, and independent mastering effective comprehensive learning materials can be used. Interactive electronic study materials in the textbook to the students with the method of science, psychological and pedagogical aspects of modern information technologies, audio and video animated proper use of the opportunities.

The edition of the electronic version of the textbook learning materials in electronic form tab is simple. However, in the form of electronic textbooks in the computer memory or other foreign elta to keep you comfortable, quick change and are characterized by the presence of to transfer over long distances via email.

Modern electronic educational materials, providing new concepts in the development of toolsfor oil and gas formation is formed. Below is the description of the implementation of the research objectives, which is optimal concepts come stir.

Elta electronic student information – storage and transfer of information in digital form a special device (in disket, CD-drive, flash drive and others).

The electronic textbook is similar to the traditional textbook chapters of different areas of science education.

Module – electronic textbook- or similar to traditional textbooks on the subject of the paragraph, department of educational sciences of the components.

Virtual laboratory of the stand –the computer laboratory of the stand-imitasiya model.

Therefore, the educational materials provided in the e-textbooks with wide opportunities of particular importance in relation to traditional publications [56,177-b.; 92, 204-209-b.; 99, 183-199-b.; 100,159-170-b.; 121,141-142-b.; 122, 74-76-b.].

Through the use of electronic textbooks and computer technology qonuniyat important concepts and mastering them opportunities to understand the process of acceleratedemergency increases.

As a result of our research work on "biology" manuals and online tutorials are created, named, in practice it has been applied.

§2.3. Supply improvement methods to read the science of biology

Preparation of highly qualified specialists with higher education at the level of the democratic developed countries on the basis of the following conditions should be done.

The development of science, this is the activities of specialists with higher education. Therefore, the training of specialists with higher education, talent is extremely modern information oqimi adopters, and individual research activities mustaqil work skills development, scientific-technical and educational information - work with the scientific literature should give you the chance to get creative.

As it is known, through the introduction of modern teaching technologies in the educational process and the specialist requirements only shots you can achieve. Read dasturli in social science-methods and theoretical knowledge in the system plays an important role. The characters, which is especially relevant to students in science, laws, concepts, models, and take in some algorithm mastering the theoretical knowledge about the importance of the role of literature in the accumulation of information is great. In general, the available scientific-theoretical and practical importance of education is clearly the problem will be solved on the basis of data method is selected. Selected teaching methods work, and the quality and efficiency of education is controlled.

Today, according to the means of static and dynamic teaching proeksion such equipment can enter the device model. Proeksion this problem using the equipment for static hal to make tryingto reflect the dynamics of the process as well because they have the possibility of lyddaq. Dynamic projections of equipment options-top: educational kinofilm in multimei brand, computer screen in the eyes with an invisible process and hsimplest show, with tools qonuniyat multiplikasiya of the process and physical-chemical mohiyat of theaudit can be. However, methodologicala perspectiveIkinofragmentlar reports from many serious studies is limited and the use of computer technology in the process of it.

Educational film, the sameas themo, multimedia (multiplikasiya) conditional on the specific level, specific to the color information, usually, ortitheemosional at cha, Ihave kkal. Image qat'iy was yesterday on the program, the dynamic speakers to toe,countryliq bthislmas, describe his methods in place to force the screen to be adapted to the image content. The image is to show a student's diqqati and directs him outlines and summerIshda known as thetu iyinchiliklarcountrywill dirib. Speakers in the audience, along withtheinterrupted asi qeha timo have, it requires special measures to restore. Educational films, multimedia, a visual and other means (devices)denying the role of qwithout ilma, in lectures on the science of the cycle we umumilmiy abstrakt fundamental interaction "revive"let's optimal means we considered the model of the device. Learning methods were the use of these tools in the training process, during our research, that is, the lessons were used .

The main reason for this activity and the implementation of innovative technologies will be increased in the interest of student teaching the lesson [60].

Innovative technologies "brainstorming" and "Pinbord" methods for the implementation of the initial lessons in the transition and independent education lesson in the effectiveness of the results of the survey raised an important factor in the implementation of it.

"Brainstorming" techniques of training of students the opportunity to work with him and allowing active participation in the process and inspiration stimulating methods.

"Pinboard" on the positive side of the main methods, that is, to work with students transported in textbooks and training manuals of chemistry is a big help to the formation of skills and qualifications, it performs independent

When a chemical experiment and have the opportunity to use information technology, chemistry for teaching a variety of topics, it is recommended to the following conclusions:

- students from using training tool to help computer similar to the experience with their own eyes for termination of chemical processes in chemical knowledge to bring more high imagine seeing tafakkurlash development;

- from the help of computer tools in the process of training of students using the structure of complex compounds multiplikasiyasiga interest is high.

The use of modern computer technics paid much attention in the educational process. All higher education institutions in the computer technique if available, the chance to use them in the process of the lesson has been increased. The use of such techniques and technologies in the educational process and increases the quality of the students is the reason for your interest. The computer used in the lesson that one should not forget that the mean does not mean that the student increases knowledge, but it is necessary to use a reasonable. Many electronic devices used in the process of education is currently very prepared. Modern electronic devices are programmed. Central to the policy for the use of information technology facilitates training in biology to learn. This increases the level of learning students are mastering the subject. In addition, this process of mastering depending on the level of students to teachers, students to teach: helps in the formation of additional questions on the topic.

CONCLUSIONS ON THE SECOND CHAPTER

1. In higher education institutions "Biology" on the subject of theoretical knowledge, laboratory work, and a demonstrated experience, independent of education achieve effective factors was recorded.

2. In higher education institutions "Biology" laboratory work on the subject of the show experience, and natural methods of organization and conduct independent virtual form without integrasiyalashgan developed and has been implemented in practice.

3. In higher education institutions "biology" study guide for science, electronic textbooks developed and has been introduced into practice (ministry of education, higher and secondary specialized October 14 89-03-3945 the year 2019-reference number).

4. In the educational process in higher education institutions "biology" modular technology was introduced to the practice of science education.

5. The constant increase of activity for chemistry teachers ' professional preparation methods, interactive methods of teaching related to at the time of their theoretical and practical knowledge, skills, and professional skills to be applied in practice through improving the efficiency of its confirmation in the research process that may serve to provide kompetentlik found.

6. In higher education institutions "Biology" science modules were developed and the constant improvement of methods of research that may serve the preparation of chemistry teachers in the process of clarification.

7. In higher education institutions "Biology" on the teaching of science based on state educational standards and educational programs to develop recommendations on the implementation of the innovative approach has been developed.

8. In higher education institutions "Biology" science is put on the content and quality of educational and innovative approaches on the basis of which state qualification requirements of the standard has been improved.

The list of used literature:

1. The republic of uzbekistan "on education"of the law
// generation – is the foundation of the development of uzbekistan. – Tashkent: East, 1997. – B. 20-29.

2. The national training program of the republic of uzbekistan
// generation – is the foundation of the development of uzbekistan. – Tashkent: East, 1997. – B. 31-61.

3. 19 February 2018 the year of the president of the republic of uzbekistan "on measures for further improvement of the sector, information technologies and kommunikasiyalari on" UP-5349-the number of times.

4. The president of the republic of uzbekistan of February 7, 2017. in "strategies of action for the further development of the republic of uzbekistan"on UP-4947-the number of times the // of the legislation of the republic of uzbekistan the package. - 2017 . 6-the number, a 70-item, 20-count, article 354, 23 number, article 448.

5. The president of the republic of uzbekistan on 29 April 2019 the year of "the system of national education of the republic of uzbekistan on approval of the development until the year 2030 kontsepsiyasini on"UP-5712-decree of the number // of law national data base. 06/19/5712/3034 number 29.04.2019 y.

6. I Karimov.A. The main task – the further development of the country and welfare of our people is in development. – Tashkent: Uzbekistan, 2010. – b. 80.

7. I Karimov.A. A free and prosperous Country, the ultimate goal of a free and prosperous life. – Tashkent: Uzbekistan, 2000. – b. 525.

8. I Karimov.A. Man, his rights and freedoms – higher values.
– Tashkent: Uzbekistan, 2006. – b. 280.

9. I Karimov.A. Of peace and security in our own strength and our power, our will and our understanding of strictly depends on. - Tashkent: Uzbekistan, 2004. – b. 400.

10. Assign Sh.M. Critical analysis strictly order-discipline and personal responsibility – the daily rule of every business leader should be. In 2016 our country's socio-economic development of the most important priorities of economic

program and the main results of the year 2017 is dedicated to. The extended meeting of the cabinet of ministers report, January 14 2017.. – Tashkent: "Uzbekistan", 2017. – b. 104.

11. Assign Sh.M. The free and prosperous, together we will build a democratic state. During the inaugural ceremony of the president of the republic of uzbekistan oliy majlis at the joint session of the chambers of / Sh.M.To assign. – Tashkent: "Uzbekistan" NMIU, 2016. – b. 56.

12. Assign Sh.M. We will build our great future together and our nation with courage and noble. –Tashkent: "Uzbekistan" NMIU, 2017. – b. 488.

13. Assign Sh.M. The development of the country and the rule of law and guarantee the interests of the welfare of the people man. Finding the main factors of 24 anniversary of the constitution of the republic of uzbekistan, dedicated to the opening ceremony part. December 7 2016 // - Tashkent: "Uzbekistan", NMIU, 2017. – b. 48.

14. Assign Sh.M. Report on the united nations.

15. A Abduqodirova.A. A Pardaev.X. The theory and practice of distance learning. – Tashkent: Fan, 2009. – b. 145.

16. Abdullayeva N. Different forms of teaching of chemistry for the use of advanced technologies and methods. - Urgench, 2011.

17. Abdullaev , Sh.V., D Muxitdinov.X. "The teaching of chemistry from the"manuals. Namangan, 2003.

18. Anvarov In N.A. College chemistry teaching a lesson on the basis of computer technologies. Dis. ... a problem. pad. nauka. – Tashkent: secondary, and vocational education of staff training and re-training institute, 2007. – b. 155.

19. Ajieva M.B. Xim obucheniya na osnove ispolzovaniya I will sovershenstvovanie soderjaniya methodical matter mestnogo the secret city (primera nor of tem "of galogen"): Dis. ... a problem. pad. nauka.– Tashkent: TGPU im. Nizami, 2007. – s. 120.

20. Azizov T.A., Ibodulloeva M.I. Soedinenii s ts o kompleksnom nikotinata common urea. //Aktual nauka sovremennoy the problem. - Moscow, - 2011. No. 2, S. - 172.

21. N. Azizxo'jaeva educational technology and pedagogical skills. — Tashkent: TDP, 2003.

22. Alimova F.A. Xim effektivnosti obucheniya a matter povishenie s ispolzovaniem kompyuternix texnologiy (nor primera razdelov "Rastvori" I "Elektroliticheskaya the disotsiasiya") Dis. ... a problem. pad. nauka. – Tashkent: TGPU im. I disputel. 2009. – s. 183.

23. I Aleynikova.I. A Savenkov.A. "Neither ispolzovanie kompyuternix texnologiy urokax ximiya". Uchebnoe posobie. - M.: Ventana. -2006. s. – 45.

24. Anvarov In N. Computer technology in chemistry // education. – Tashkent, 2003. – № 3. – B. 7-9.

25. Anvarov In N. Pedagogical possibilities for the use of computer software for the teaching of chemistry. A collection of articles o'zpfiti. – Tashkent, 2007. – b. 225 - 227.

26. Ayasi A. A Demirvas. Turkish second introductory chemistry concepts visionary conceptions of students // journal of Chemical increasing. 1997. – r. 745. – r. 518-521.

27. Axmerov Q., All The A, R Sayfuddinov. General and inorganic chemistry - Tashkent: "Uzbekistan", 2003.

28.Bahrieyv A. n. through the formation of new pedagogical technologies of teaching bahrieyva possessing internal motivation. / "Public education" in the journal, 2006, № 6. - b. - 25-28.

29. It Is Begimkulov.Sh. and others. Pedagogical education: theory and practice. Monographs. – Tashkent: Fan, 2011. - 232 b.

30. Begmatov Sh.Sh. Urokov zanyatiy po osnovi vzaimosvyazi I metodicheskie vneurochnix a matter xim: Avtoref. dis. ... a problem. pad. nauka. – Tashkent: Uzbek nauchno-issled. the institute of the pad. nauka im. T.N. Pecan-Niyazova. 2003. – s. - 17.

31. K Belikov.B.Formirovanie professionalnoy kompetentsii budutshix teachers matter xim po razvitiyu uchatshixsya kreativnosti. Dis not. so neither. three. st. what questions. pad. nauka. Effect. 2008.

32. Belov P. S. Formirovanie ximicheskix kompetentsiy obuchaemix neither prakticheskix zanyatiyax po xim matter. Dis not.so neither. three. st. what questions. pad.nauka. Nizhny Novgorod. 2012.

33. BespalkoъV ko.P. Progerssivnie obucheniya I texnologii pedagogy.
- M.: Izda-where professionalnogo obrazovaniya in the institute Of obrazovaniya Rossi, 1995.

34. Bespalko. V. P. Kompyuterov I obrazovanie obuchenie s uchastiem (pedagogical tretego tisyacheletiya). – M: Izda-where Moskovskogo psixologo-sosialnogo in the institute. 2002. – s. 352.

35. Berdiqulov R. Sh., Iskandarov O. Yu., Mirkomilov Sh.M. Inorganic chemistry. - Tashkent: "Thought bo'stoni"publishing house, 2018. – B. 48-56.

36. Berezin B.D. T.N.Lomova. Reaktsii dissosiasii kompleksnix soedineniy. - M.: "Nauka", 2007. - s. 278.

37. Bodi features.O.G. Virtualnaya laboratory — odno sredstv obucheniya intensifikasii prosess xim matter in track. http://him.1september.ru/articlef.php

38. The Observer N. I. General chemistry. Textbooks. – Tashkent: TDP, 2019. b. 318-338.

39. The Observer N. I. Ibodulloeva and M.I. "Biology" educational-methodical manual, - Tashkent: TDP, 2015. – B. 98.

40. Borovskix T. A. Propaedeutics metodicheskoy podgotovki budutshix na pervom xim was pedvuza teachers matter. Avtoref. dis. ... a problem. pad. nauka. Focusing sp, 1998.

41. Vasilyeva P.D. Professional-metodicheskaya podgotovka v vuze city xim teachers matter: Sinergeticheskiy podxod: Monographs. - Sp malli: Izda-where RGPU im. A.I. Gersena, 2003. – s. 193.

42. Vasilyeva P.D., Kuznesova N.E. Xim obuchenie matter. - Sp malli: TILE, 2003. - s. 128

43. Verbiskaya L.A. Problem sovremennogo universitetskogo obrazovaniya // universitetskogo obrazovaniya v usloviyax sovremennix Modernization. Vip. 2. -He sp: SPand bg,2001. -S. 5-11

44. Gavronskaya Yu.Yu. Kompetentnostniy interaktivnogo obucheniya studentov v pedagogicheskom vuze podxod k proektirovaniyu ximicheskim dissiplinam. // Sibirskiy pedagogicheskiy journal. - 2008. - 8. - s. 47.

45. Gabrielyan O.S. V uchebnom sovremennom information is prosess / O.S. Gabrielyan, S.A. Sladkov // Ximiya v shkole. – 2008. - № 7. S. – 23.

46. L Golish.V. Active teaching methods: content selection, implementation. express guide. -T: TASIS, 2001.

47. L Golish.V., D Fayzullaev.N. Planirovanie I organizasiya pedagogicheskix texnologiy. – Tashkent: "Economy", - 2010. s. 45-47

48. Sh Daminov.Sh. Tour N.N., Aliyorov S.K. Laboratory sessions are inorganic. – T: Uzbekistan, 2006. -96 - b.

49. Dejan dinevski, Fošnarič Heaven, in tanja arh. Open educational resource in e-learning. – *Journal of Chemical Education.* Voles.43. no.1. 2010. p.4-8.

50. E Has Been Fertilized.K. Xim texnologii v obuchenii I sovremennie innovasii matter. - Kaliningrad - s. 2000.-18-44.

51. M. o. Yoriyev and b. Issues are general and inorganic and exercise pack. Training guide. Tashkent. 2008.

52. Jarkix L. A. Xim po razvitiyu umeniy ustanavlivat deyatelnost city teachers a matter prichinno-svyaz sledstvennie vospitatelnix zada president when the pride of reshenii: Avtoref. dis. ... a problem. pad. nauka. Focusing sp, 2004.

53. Jodi L. Seas, gae leinhardt, james Green, Kennith Koedinger, David Klahr, Michael Karabinos, and working around ideas D., "evidence-based Approaches to improving instruction in chemical equilibrium". *Journal of Chemical Education*, Voles. 95. 2014. - 10, pp. 1517-1525.

54. Zaytsev O.S. "In a matter xim obucheniya methods: Teoriticheskiy I prikladnoy aspects of" Moscow: izdatelskiy further humanitarian cents "By", 1999. s. 98-144.

55. Zaylobov L.T. General chemistry the oxidation of kaytarilish is to read with the help of information technology improvement (akadem lyceums): Avtoref. dis. ... a problem. pad. nauka. – Bukhara, 2012.
56. MWF zenkina. Computer software obrazovaniya studentov kak sredstvo uchebnogo naznacheniya povisheniya effektivnosti ximicheskogo estestvennonauchnix spesialnostey. Dis not. what questions. pad. nauka. - Stavropol, 2001. - s.177.
57. Zlotnikov E.G. Experiment methods obucheniya kak ximicheskiy spesificheskiy //Ximiya. Moscow: September pervoe city.- 2007. №24. - s. 18-25.
58. Inoyatov Is It.I. Muslimov N.A. D Ro'ziev.I. M.H.Usmonboeva. Pedagogical. Textbooks. Tashkent: TDP. 2013. - B. 256.
59. A Iskandarov.Yu. Issues are methods to take off. Textbooks.-Tashkent: TDP, 2018 - b. 220.
60. Ishmuxamedov R.J. Innovation in educationn technologies.
// The journal of continuing education. 2008. 5 - number.
61. Karimova D, E. the formation of organic chemistry basic concepts of Educational Jumayeva. // The journal of pedagogy. Tashkent. 2016. 6 – number. - b. 76 - 83.
62. Kostromina N.A. V Kum Was Over.N., Skorik N.A. Ximiya koordinasionnix soedineniy. - M.: Visshaya shkola, 1990. – s. 432.
63. Kuznesova N.V. "In a matter prepodavaniya methods xim". – Moscow: izdatelskiy further humanitarian cents "by the" 1999. s. - 95-117.
64. Kukushkin Yu.N. Ximiya koordinasionnix soedineniy. M: Visshaya shkola, 1985. - 455 p.
65. Krasnova O.M. Romanov In O.V. Metodicheskoe proektirovanie I konstruirovanie prosess in a matter xim obucheniya: information texnologii v obuchenii xim matter. Modulnaya software. - Rostov n/d, 2006. – s. 48.
66. Mamadaliyeva A., M. Ibodulloyeva case study organic chemistry in the laboratory sessions and assignments on the basis of organization. // The journal of pedagogy. Tashkent. 2017. – 6 - the number of. B. – 48 - 51.

67. Mamajanov Sh.A. Higher education institutions "chemistry and its teaching method," the structure and content of the stage.: Avtoref. dis. ... a problem. pad. nauka. – Fergana: and fd, 2003. – b. 22.
68. Mirkamilov Sh.M. I Shernazarov.E. Academic lyceums "organic chemistry," stage "hydrocarbon" is the subject of the introduction of modern information technology to read // Pedagogical education. - Tashkent, 2006. 5 - number. B. - 42 - 44
69. Metodicheskoe ukazanie - "Ximiya koordinasionnix soedineniy", - Tashkent: Tash, 1987.
70. A Muraveva. A. Kuznesova Yu., Chervyakova T. N. Organizasiya modulnogo obucheniya, osnovannogo neither kompetentsiyax: posobie prepodavateley beginning. — M.: Alpha. 2005.
71. Nizamov S.A. Innovative technologies for the teaching of chemistry to be applied // Pedagogical education. – Tashkent, 2012. –№3. B. 69-73
72. Nizamov S. A. Innovative technologies of raising the effectiveness of chemical education. Avtoref. dis. ... candidate of philosophical sciences. – The national university of uzbekistan, 2018. – b - 24.
73. Nishonov M. Mamajanov Sh.A. Institutions of higher education "teaching of chemistry" improving the structure and content of the phase. // Pedagogical education. – Tashkent, 2004. 2 - the number. B. 52-53.
74. Nishonov M. Mamajanov Sh., Hodjaev V. "The teaching of chemistry". T.: "The Teacher". 2002. - b.111.
75. It Is Nishonaliev.N. Education and pedagogical standards innovasiyalar – T; // education. 1999. 6 - number. B. 28 – 31.
76. Buyeynikova O. N. Texnologii module: razrabotka I proektirovanie I have obrazovatelnix program: uchebnoe posobie. - M.: Alpha -M.: INFRA-M. 2010. – s. 256.

77. Omonov H.T. Filosofsko-pedagogicheskie osnovi ego sovershenstvovaniya ximicheskogo voprosi I obrazovaniya: Avtoref. dis. ... dok. pad. nauka. – Tashkent: TGPI im. Nizami, 1995. – s. 46.
78. Omonov H.T., Library. H Teshaboev. S Nishonov M. Tulaganov A. Chemistry. // The development of education. – Tashkent, 1999. – the number of. – B.
79. T. omonov h., n. h. Tleuov, Madyarova S. A, ua eshchonov e. u. pedagogical technologies and pedagogical skills. Tashkent. "Economy and finance". 2009, - b. 240.
80. New pedagogical technologies S. Og'ayev-life needs. // "Public education" in the journal, 2001, № 3. – b. 69-71.
81. Pak M.S. "I obucheniya xim any methods in the theory of matter". Saint-Petersburg: Izdatelstvo RGPU im. A.I.Gersena, 2015. - s. 249-279.
82. Paink m.S. Methodology issledovaniy v nauchnix oblast ximicheskogo weather-razovaniya // perspektiv razvitiya obrazovaniya I will ximicheskogo problem: fast. at dok. II Vseros. nauch-konf praktika. - Chelyabinsk: CHGPU, 2006. - s. 24 - 27.
83. Parp N.A. Rakhimov X.R., A Muftaxov.G. Theoretical basis of inorganic chemistry. - Tashkent: "Uzbekistan", 2000.
84. Parp N.A. Inorganic chemistry. -Tashkent: Uzbekistan, 2003. -504 b.
85. Parp N.A. A Muftaxov.G., Rakhimov X.R. Inorganic chemistry. Tashkent Uzbekistan, 2003.
86. Parp N.A. Umarov B.B. "Ftorli R - diketon surge of people and their complex combinations". Tashkent: "Design - press". 2012.
87. Pedagogy: Uchebnoe posobie. // V.A.Slastenin, A.I.I mitshenko dr. – 3-e izda., M.: Shkola-press, 2000 – s. 512.
88. Polosi B. C., Prokopenko V.G. Praktikum po xim is methodical prepodavaniya matter. Uche. posobie beginning studentov pad. in-tov. — M.: Prosvetshenie. 1989. - 224 s.
89. Rajabov X.M. Conceptual and historical aspects of teaching organic chemistry (chemistry of the direction of education of the higher educational institutions in the example) PhD abstracts, republic of nukus, 2019, b. - 8.

90. RahN matullaev.G., Omonov H.T., Mirkomilov Sh. M. Chemical arrow onyour dogsh methods. - Toshkent: "economy and finance", 2013. - B. 320.
91. Raymond Powder. General Chemistry: The Essential Concepc. 5 edition, England, 2013. - r. 699.
92. Romanov In O.V. Professional o system-this metodicheskoy podgotovkik ispolzovaniyu xim city dutshego teachers matter informasionnix communication Inv obuchenii texnologiy ix // Spesvipusk "Pedagogy". I humanitarian sosial-ekonomicheskie nauka, 2006. №2. S. 204 - 209.
93. D Ro'ziev.I. Usmonboeva M.H. Interactive methods: significance and can be applied. Methodical guide. Tashkent:. TDP. 2014. – B. 113
94. Salkova T.P. Pedagogicheskie texnologii: uchebnoe posobie. M: TT Sfera, 2005. s. 44-46.
95. Safarova R. G., Yusupova F. I. Sulaymonova G. T., Ko'chkarova F. M. Atakulova N. A. Karaxonova O.Yu. The technological process of ensuring modular education school of theoretical-methodological basis. Tashkent: Cho'lpon. 2015
96. CG elevko.K. Sovremennie obrazovatelnie texnologii.–M.: Narodnoe obrazovanie, 1998. – s. 256.
97. Skopenko V.V., Tsivadze A.Yu., Savranskiy L.I. A Garnovskiy.D. Koordinasionnaya ximiya. - M.: The oic "Akademkniga", 2007. - s. 487.
98. Steshina O.S. Metodicheskie osnovi rezultatov nauchnix issledovaniy redoks primeneniya-v shkole v prepodavanii sistem xim visshey matter. Avtoref. dis. ... a problem. pad. nauka. – Tashkent: TGPU im. Nizami, 2006. – s. 21.
99. The Flavors Of A Z. A Ayasi. Turkey virtual chemistry laboratory: effects of constructivist learning environment // Turkish online journal of distance education-the crown., 2012. 13(1). – r. 183-199.
100. The Flavors Of A Z. A Ayasi. Virtual Chemistry Lavoratory effects on Students' Ashievement // Teshnology educational & Society, 2013. 16(1). 159-170.
101. Tolipova J.O. Innovative teaching technologies policy central to biology. Textbook for students of pedagogical universities. – T: "Cho'lpon", 2011.

102. Toshpulatov Yu.T., Ishokov Sh.S. Inorganic chemistry. Tashkent, "Teacher", 1992.

103. Toshpulatov Yu.T., Rakhmatullaev N.G. Theoretical basis of inorganic chemistry. - Tashkent: "The Teacher". 2005.

104. Tüysüz C. The effects of the virtual laboratory on students' achievement and attitude in chemistry reflects. The international online journal of educational Science, 2 (1). 2010. - r 37-53

105. Was Awarded With R., N. A. Sh. General theoretical and practical training. - Tashkent: Uzbekistan, 1997. – b. 192.

106. Uchebnoe elektronnoe izda: Ximiya. 8-11 class. Virtualnaya laboratory // Laboratory sistem multimedia. – Marg, - 2004.

107. Elektronnoe uchebnoe izda: Virtualьstem in the laboratory. (2CD + metodicheskoe posobie.) // Mul laboratory sistemьtime. Marg, - 2006. In: 1000 MV. – 1000 S.

108. A Xayitov. Kontsepsiyasi the computerization of secondary general education // education. – Tashkent, 2003.2 - the number of. – B. 93-95.

109. Chernobelskaya G.M. Xim obucheniya v sredney shkole methods matter. Moscow: "By", 2000. – s. 127-153

FSC
www.fsc.org
MIX
Papier | Fördert
gute Waldnutzung
FSC® C083411

Kolibri 360 GmbH
Ferdinand-Jühlke-Straße 7
99095 Erfurt, Deutschland
produktsicherheit@kolibri360.de

Druck:
CPI Druckdienstleistungen GmbH
im Auftrag der
Kolibri 360 GmbH
Ferdinand-Jühlke-Str. 7
99095 Erfurt